MEMBRANE MICRODOMAIN REGULATION OF NEURON SIGNALING

Membrane Microdomain Regulation of Neuron Signaling

Ron Wallace

Nova Biomedical Books
New York

Library of Congress Cataloging-in-Publication Data

Wallace, Ron, Dr.
Membrane microdomain regulation of neuron signaling / Ron Wallace.
p. ; cm.
Includes bibliographical references and index.
ISBN 978-1-60456-318-4 (hardcover)
1. Neurons. 2. Cell membranes. I. Title.
[DNLM: 1. Neurons--physiology. 2. Membrane Microdomains--physiology. 3. Nervous System Diseases--physiopathology. 4. Synaptic Transmission. WL 102.5 W193m 2008]
QP363.W35 2008
612.8'2--dc22 2007051509

Published by Nova Science Publishers, Inc. ✢ New York

Contents

Preface

The rapid pace of discovery in membrane molecular biology is revealing unexpected complexity at the boundary of the cell. The membrane appears to be more than a thin film separating aqueous compartments or an anchoring site for proteins. Indeed, it shows many indications of being a dynamic, regulatory structure composed of transient lipid ensembles known as *rafts* or *microdomains*. These appear to regulate the membrane-protein kinetics which are the basis of many cellular functions. This book describes those regulatory features in detail and presents evidence that they operate in the neuron. An unorthodox implication of the model is that the action potential, or nerve impulse, is regulated by an immense number of neural-membrane molecular systems. If correct, this would mean that the neuron is not the fundamental unit of brain information-processing. Rather, it may be viewed as a linked series of molecular-computational systems that systemically regulate an overall pattern of excitability. Thus the book may be of interest to researchers in computer science and nanotechnology. In particular, it should appeal to investigators designing biomimetic devices. Also, the present approach is not without medical implications. Drug design in neuroscience has largely overlooked the membrane, concentrating instead on ion channels, transmitters, receptor sites, and genetic therapies. Although these approaches have led to highly useful etiological models and treatments, the recent findings in membrane biophysics are a plausible source of additional therapies.

My involvement with this topic happened in an indirect way, more or less as the result of one question leading to another. I am trained as an anthropologist with a specialization in human evolution. For several years I was interested in the neurobiology of language and what the cortical and subcortical systems subserving verbal behavior might tell us about language evolution. Specifically, I was persuaded (and still am) that human spatial mapping and language originated and evolved together. A subsequent research sabbatical at the Netherlands Institute of Brain Research, where I conducted a project in rat hippocampal developmental neuroanatomy, provided me with valuable laboratory experience relevant to spatial mapping and an opportunity to discuss with specialists a wide range of topics in the neurosciences.

I returned to the States at about the time that a somewhat unusual theoretical debate was emerging in brain studies. A relatively small number of researchers in biophysics and other disciplines was proposing that molecular systems in the neural membrane or the cytoskeleton might regulate neuron signaling. Almost immediately this debate became linked to philosophical speculations, including unresolved questions in the interpretation of quantum mechanics. These discussions became highly speculative, including some viewpoints that (in my opinion) were not testable even in principle. Of much greater interest to me were the neuromolecular-computing models, shorn of the philosophical arguments. Because of my background in evolutionary biology, it seemed to me that the molecular computing systems, if they in fact existed, might vary between species (and between brain structures within a species) in their computational power. A programmatic article in which my biochemistry colleague Harry Price and I examined this possibility and suggested possible experimental approaches appeared in *Biological Cybernetics*. Much of that article's biophysics and some of the experimental discussion have found their way into the present book. This work, however, does not discuss evolution---with the important exception of the research strategies briefly suggested in the conclusion. Although the evolutionary aspect of molecular computing does continue to interest me, my primary concern here is to identify the inner workings of neural membrane microdomains, and their possible implications for the etiology of neural disease.

I hope that this book will be of interest not only to researchers in biophysics and neurobiology, but also to those students (especially in cell biology) who are interested in neuron function. Thus while certain assumptions have been made regarding the reader's background, some concessions have also been made to invite a wider audience. It is assumed that the reader understands the essential molecular and ionic processes of the neural impulse (the Hodgkin-Huxley action potential). Although these are reviewed briefly when introducing the model, there is no detailed discussion resembling that found in a standard textbook.

Concessions have primarily been made in the mathematical discussions. For the most part, this book is a qualitative study of a specific domain in cell biology and its medical implications. However it has been neither possible nor desirable to completely exclude mathematics. Biophysicists rightfully require a high level of quantification as a basis for evaluating a model's internal consistency as well as its compatibility with related models at different orders of magnitude. Put less formally, quantification can (usually) indicate if a model is physically plausible, and if it is consistent with macroscopic and microscopic physical contexts. For that reason, the explanation of the model in Chapter 3 utilizes the mathematics---mostly differential equations---necessary to accurately describe the response of a molecular system to an applied electric field. But in the interests of a larger audience---and at the risk of some redundancy---the equations are followed in all cases by qualitative explanations. This approach may have a benefit in addition to accessibility. It helps insure that each quantitative step of the argument is clearly linked to a biophysical process. The mathematics has thus not been allowed to take on a life of its own separate from its biological referents.

Without the patient assistance of my colleagues in chemistry and physics, this book would not have been possible. When I first began work in this area, I benefited from frequent discussions of quantum mechanics and molecular information-processing with Ralph Llewellyn and Michael Johnson of the University of Central Florida Department of Physics and Binayak Dutta-Roy of the Saha Institute of Nuclear Physics, University of Calcutta. During a later trip to Finland, my wife and I were graciously hosted by Paavo Kinnunen, Department of Medical Chemistry, University of Helsinki who first pointed out to me the significance of unsaturated lipids in neural membrane regulatory processes. As the model grew more detailed, it was critiqued by Ole Mouritsen of the Department of Physical Chemistry, Technical University of Denmark, Alwyn Scott (now retired) formerly of the Department of Mathematics, University of Arizona and Michael Conrad, who until his untimely death was a pioneering researcher in biomolecular computing theory at the Department of Computer Science, Wayne State University. I am also grateful for constructive criticism from participants in the Eighth Foresight Conference on Molecular Nanotechnology and the 42nd and 43rd Sanibel Symposia on Atomic, Biophysical, and Condensed Matter Theory.

I am especially indebted to my colleague and friend Harry Price of the Department of Chemistry, Stetson University. To a large extent this book is a result of his efforts as much as my own. We began collaborating on this problem in the summer of 1996, and have continued to work on it until the present. He is responsible for the computational modeling of membrane molecular processes, and developed several important aspects of the theory. It was Harry who first recognized that membrane lipid ethenes aligned in the plane of the bilayer are polarized by an applied electromagnetic field. That aspect of the model then led us to consider electrostatic interactions between membrane dipoles and charged amino-acid residues in a gated ion channel. In addition to these specific contributions, I am grateful for his openness to interdisciplinary dialogue and his willingness to consider unconventional ideas. I am certain that my own understanding of membrane biophysics would not have been possible without his judicious combination of patience, clarity, and rigor.

Above all, I am grateful to my wife Susan for her continued support. She assumed more than her fair share of household and community responsibilities, including caring for our aging and occasionally ill-tempered samoyed Sammy, so that I could work on my writing.

Ron Wallace
Orlando, Florida

Chapter I

Introduction

The thin film of molecules that envelops the living cell is revealing itself as a structure of unexpectedly subtle complexity. Once dismissed as a simple barrier and an anchoring site for proteins, the membrane for the last 30 years has been widely recognized as an internally organized system essential to many biological functions (Jain, 1972; Alfsen, 1989; Gennis, 1989; Kinnunen, 1991; Mouritsen and Jørgensen, 1992; Mouritsen and Jørgensen, 1994; Jørgensen and Mouritsen, 1995; Lehtonen et al., 1996; Dumas et al., 1997; Lehtonen and Kinnunen, 1997; Simons and Ikonen, 1997). Membranes contain discrete regions in which energy transduction mechanisms such as photoreception are organized. Other membrane sites are dedicated to enzymatic reactions, active and passive transport, ion channel activity and receptor site function. The high level of organization, like that of all other biological structures at many different orders of magnitude, is the outcome of natural selection. During 3.5 billion years of cellular evolution, the membrane has functioned as a boundary-maintenance system, mediating molecular interactions between the cytosol and the extracellular fluid. If extracellular and cytosolic traffic had been totally regularized, the membrane would have been a structurally simple "look-up" system matching invariant biochemical inputs to invariant responses. But the actual membrane milieu has probably never been static. Both membrane surfaces have been continually affected by changing levels of electrolytes, metabolites, transmitters, cations, and phosphorylation-driven intermolecular associations. Thus the molecular input structure---species and concentrations---has been combinatorially explosive, generating high complexity at the edge of the living cell.

The present book argues that the complexity of the membrane---in particular its internally compartmentalized, regulatory features---permit it to function in neurons as a computational system which regulates ion channel dynamics. In essence, *the neural membrane regulates neuron signaling.* In this chapter the argument will be outlined, and in subsequent chapters it will be explored in detail. Although the book's primary purpose will be to defend a novel view of nerve cell information-processing, medical aspects of the model will be examined as well. Regarding that objective, a cautionary note is appropriate. The book is based on studies of natural and artificial membranes and computer simulations of membrane-lipid and lipid-protein dynamics. The findings are highly convergent in their identification of biophysical properties which are also characteristic of nerve cells. However, at the time of writing, few direct studies have been done of neural membrane interaction with ion-channel proteins. Thus this book is simultaneously a synthesis of recent research and a call for more extensive studies. If the reader is uncomfortable with programmatic approaches, this book is probably not for her or him. Caveat lector.

Viewed in historical perspective, the model developed here is one of many recent contributions to a lengthy and spirited debate about membrane molecular biology. Chapter 2 is an interpretive historical summary of major theoretical and methodological contributions to the study of membrane structure and function. It begins with the pioneering 19th-century work of Charles Overton and concludes with recent simulation approaches and investigations of natural and artificial membranes. A theme in this history is the increasing evidence of molecular cooperativity in the membrane, and the difficulty of finding a satisfactory explanatory model (Tanford, 1989). In this context, the Fluid Mosaic Model of S. J. Singer and Garth L. Nicolson (1972) is examined in some detail. It is suggested that, although this model has been modified substantially since its introduction, its emphasis upon regions of transient molecular order surrounded by a patternless "two-dimensional" liquid was a major conceptual advance in the understanding of membrane dynamics.

In Chapter 3 the model is presented. Its basic mechanism is field-induced electrostatic interactions between charged amino-acid residues in an unfolded ion-channel protein and dipoles in the surrounding membrane crystalline structure or *microdomain* (Poo and Robinson, 1977; Lee et al., 1994; Papazian et al., 1995; Tiwari –Woodruff et al., 1997; Groves et al., 1997; Radhakrishnan and McConnell, 2000; Chen et al., 2000; Price and Wallace, 2001; Torshin and Harrison, 2001). These charge-charge interactions are proposed to regulate the mean open time of the channel. As a result, they ultimately regulate the propagation of the electrical signal. This electrophoretic approach is believed to represent a major conceptual shift regarding the ultimate unit of brain information-processing. Since the formulation of the Hodgkin-Huxley (HH) model of the action potential (Hodgkin and Huxley, 1952a, 1952b, 1952c, 1952d) and the development of related artificial-intelligence constructs such as the McCullough-Pitts neuron (McCullough and Pitts, 1943) the nerve cell has been identified as the brain's fundamental informational unit (Churchland and Sejnowski, 1992). The evidence presented in this chapter suggests that this viewpoint may not be correct. Although it is indeed frequently useful to think of the neuron as a switch, it is probably more accurate to view it as a system "composed of many smaller switches located throughout both its dendritic and its axonal branches" (Scott, 1995: 58). Although this computational claim is examined in some detail in Chapter 4, it is first introduced here in relation to specific molecular mechanisms. In particular the role of membrane fluidity in modification of the molecular "hardware" is described at some length (Kinnunen, 1991; Simons and Ikonen, 1997).

Chapter 4 examines the possible role of neural membrane microdomains in regulating the propagation of the action potential. It begins with an overview of the experimental evidence for AP conduction failure. It then examines the two major alternative models: one emphasizing the role of impedance mismatch due to neuron branching geometry; the other emphasizing the role of prolonged hyperpolarization due to the A-current potassium channel (K_A). A model emphasizing the possible interaction of a microdomain-cytoskeleton system with neuron branching geometry is then presented.

Chapter 5 examines the possibility of designing a molecular machine that would mimic neural membrane microdomain properties. The basic architecture of the proposed device is an electrically-addressable biotemplated nanowire orthogonal to an artificial membrane. The engineering difficulties as well as the potential advantages of the device's components are discussed at length. Despite the difficulties involved, it is suggested that a device of this type could be useful in modeling the etiology of neural diseases. The possibility is illustrated with reference to the A-current potassium channel which is believed to play a significant role in epilepsy.

Chapter 6 summarizes the argument, and suggests additional lines of research not explored in the previous chapters. Because I am an anthropologist, evolutionary approaches to microdomain dynamics appear especially promising (Wallace, 1995, 1996; Wallace and Price, 1999). The idea is not entirely novel. There is already a significant data set in comparative molecular neurobiology, and some important work has been done in tracing the ancestry of ion channels (Harris-Warwick, 2000). But as the latter example suggests, the molecular biology of the neural membrane has largely been overlooked in evolutionary reconstructions. The concluding chapter calls for more research in this area, and outlines a sample investigation. It is concluded, optimistically, that this research direction would generate an enlarged understanding of neural evolution, as well as providing additional evidence for the “blind watchmaker” of natural selection operating at the molecular level (Dawkins, 1986).

Chapter II

Membrane Studies: The Problem of Order

The history of membrane studies can be largely understood as the recognition and delineation of an orderly structure. The rationale, evident today, that a cell requires an external semipermeable barrier to regulate the passage of substances and thereby maintain homeostasis was by no means evident to the earliest cell biologists. Thus the earliest and most fundamental scientific breakthrough---occurring in the late nineteenth century---was simply to recognize that the membrane in fact existed (Harris, 1999). For both sides of the debate, the dominant metaphor for a cellular barrier was not that of a deformable, semipermeable structure but rather a rigid wall, not unlike the cell wall in plants. This rigidity seemed incompatible (claimed the opponents of the membrane concept) with the observed mobility of unicellular animals and the ligature required in cell division. In their view, the cell was a gel-like structure that flowed to explore its environment (in the case of unicellular animals) and invaginated to divide (in both unicellular and multicellular species). The limited evidence---primarily Protozoa studies---appeared momentarily to support those who denied the membrane's existence. But the quality of the microscopic observations---and the dominant metaphor of a rigid cellular envelope--- changed dramatically within half a century.

2.1. The Globule Model and the Membrane (1800-1850)

The concept that biological membranes actually existed and functioned as selective barriers was originally proposed by French cytologists in the early nineteenth century, only to be rejected by their German counterparts. Henri Dutrochet (1776-1847), who may have formulated a "cell doctrine" independently (and possibly prior) to Theodor Schwann, argued in 1824 that the "globule" was the basic unit of metabolic exchange, its passage of nutrients and wastes regulated by a semipermeable structure (Hughes, 1989; Harris, 1999). It is unclear if Dutrochet, as well as other cytologists in the early nineteenth century, was using the term "globule" to mean what today would be called a cell or indeed if he had actually observed a membrane. It is possible that he was observing circles in the visual field created by the optical interference of his low-aperture microscope. If this was in fact the case, then the artifact probably led him to the notion of a discrete unit of life, thus exemplifying a correct idea derived from a flawed apparatus. His subsequent animal-tissue experiments in turn became the basis for arguing that the globular units must have membranes. Dutrochet reasoned that the morphological uniformity in tissue depended upon constituent units that contain and secrete a uniform substance. "It is in the cells," he claimed, "that the fluid appropriate to each organ is secreted." (Dutrochet in Harris, 1999:29). He concluded that a surrounding membrane must somehow regulate the inflow of the "appropriate" substances and expel wastes from the cell's interior. This viewpoint influenced, among others, François-Vincent Raspail (1794-1878) who proposed in 1843 that the selectivity described by Dutrochet must be due to the composition of the barrier. "Cells have varied means of choice," he argued, "resulting in different proportions of water, carbon and bases, which enter into the composition of their walls" (Raspail in Harris, 1999: 32).

2.2. The German Reaction: The Membraneless State Model (1850-1890)

The views of Dutrochet and Raspail almost immediately encountered resistance, most pronouncedly from cytologists in Germany (Harris, 1999). One of the earliest proponents of what would soon be called the "membraneless state" model of the cell was the botanist Anton de Bary, who conducted research on slime molds (the Gymnomycota). These somewhat unusual organisms, generally found on damp soil or other decaying organic matter, resemble plants at some points of their life cycle, animals at others. What most influenced de Bary---and in fact became the paradigmatic example for his membraneless state model (1860) – was the vegetative phase of the life cycle during which the slime mold develops a large multinucleate mass or plasmodium. In this animal-like phase, the amoeboid organism moves about slowly, ingesting organic matter through phagocytosis. It was evident to de Bary that the plasmodium's freedom of movement was incompatible with a "Membran", a word widely used at the time (especially among German cytologists) to connote a rigid cell wall. Second, and more important, his intensive studies of these organisms---he had investigated slime molds for many years and had written a classic monograph on the subject---indicated that cellular nuclei need not be separated by a partition. He concluded that animal cells did not require membranes, and that alterations in the shape of a cell were due to the internal contractility of the protoplasm.

De Bary's views were warmly greeted by his fellow botanist Max Schultze (1825-74). In 1858 Schultze had arrived at the same conclusion based largely on his studies of North Sea diatoms (Chrysophyta), organisms that, at first glance, would seem totally at odds with the Membraneless State Model (Harris, 1999; Hughes, 1989). With their distinctive boxlike structures and silica-impregnated glasslike walls, they appear to be anything but the "naked blob of protoplasm" which was the essence of the De Bary-Schultze viewpoint. But Schultze dismissed the rigid structure as an artifact of a hardening agent. As we have seen in the case of Dutrochet's microscope, experimental techniques and apparatus in cell biology were still relatively primitive. As a consequence, Schultze's argument carried considerable weight in its day. As he and his contemporaries knew, only three years earlier (1855) Robert Remak had described experiments with hardening agents in a pioneering study of the embryology of vertebrates. After trying a number of substances---hydrochloric, sulphuric and chromic acids, mercuric chloride, alcohol---Remak had discovered that with a reagent of 6 percent copper sulphate and an equal volume of 20-30 percent alcohol he could visualize blastomere cleavage in a frog's egg. Schultze had replicated these experiments and, like Remak, had visualized an apparent membrane. But given the state of the art, he remained highly skeptical, strongly suspecting that the observation was merely an effect of the reagent.

Schultze's argument against the membrane involved more than purely technical criticism. He was persuaded, based on evidence of cytoplasmic movements primarily obtained by Franz Unger (1800-1870) and Ferdinand Cohn (1828-1898), that the cellular contents had a contractile property. This mysterious trait enabled the cell to change its shape, an ability which would be "impeded, if not altogether rendered impossible, by the presence of a rigid cell membrane" (Schultze in Harris, 1999: 150). The idea was not novel. Claims for a "contractile substance" had been circulating among European cytologists since 1835 when Félix Dujardin formulated the idea of the "sarcode". This term was applied to the cytoplasm and meant a "glutinous, diaphanous substance that contracts into globular lumps" (Dujardin in Harris, 1999: 74). Franz Unger was convinced that he was observing Dujardin's sarcode in a squash preparation of *Malva sylvestris* pollen. In a description of cytoplasmic activity that flirts dangerously with vitalism he noted that "the movements were not oscillatory but at different times advanced, retreated, moved sideways or gyrated…This marvelous, or at least astonishing, scene resembled an army of monads *full of inner vitality, full of an inner self-determination that revealed itself in their movements*" (Unger in Harris, 1999: 110; emphasis mine). Ferdinand Cohn, in an 1850 paper, reported these movements in bacteria, concluding that the contractile substance was an essential characteristic of cells. These claims, in combination with the uncertain reliability of the early reagents used in membrane experiments, caused Schultze to conclude that cells were membraneless blobs.

Disputation over the existence of the membrane continued until the end of the nineteenth century (Harris, 1999). During that period, the Membraneless State Model was significantly weakened by the investigations of osmosis in plant cells performed by Hugo de Vries (1848-1935). The concept of cellular osmosis, although controversial, was by no means new. It had been proposed in the early nineteenth century by Dutrochet as an explanation of muscle contraction and later in the 1850s by Nathaniel Pringsheim as an explanation of plant cell plasmolysis (the retraction of cytoplasm into a smaller volume). De Vries described these and other experiments in the light of his own studies of plant cells suspended in various salt solutions, particularly potassium and calcium. His argument was essentially an appeal to parsimony. He maintained that the changes in turgor pressure observable in plant and animal cells were most readily explained by the concept of a mediating membrane, and not by a rigid cell wall. In the bargain, he was also rejecting the vitalistic "sarcode" proposed by Dujardin, Unger, and Cohn in favor of a mechanistic model. Following De Vries' publications, the weight of scientific opinion began to shift in favor of the membrane. But the final---and decisive---blow to the Membraneless State Model would be delivered by Charles Ernest Overton following a series of experiments begun in 1889.

2.3. Early Models of the Cell Membrane (1895-1925)

On October 31, 1898, Charles Ernest Overton, a botanist primarily known for his research in plant heredity, addressed a gathering of the Science Foundation in Zurich. Heredity, however, was not his topic. Instead, he presented a large body of evidence which effectively demolished the Membraneless State Model and the associated "sarcode" viewpoint which had dominated cell biology since the days of de Bary and Schultze. His introductory remarks were forceful, the modest qualification at the end appearing almost as an afterthought: "I have been carrying out investigations on the general osmotic properties of plant and animal cells for more than nine years. After having performed some 10,000 experiments with more than 500 different chemical compounds I should have reached a satisfactory general view about them" (Overton, 1899 in Branton and Park, 1968b: 45). The series of experiments to which Overton referred investigated the permeability of cells in aqueous solutions containing various types of compounds. His first studies resembled those of De Vries in their demonstration that plasmolysis could be generated by incrementally increasing the concentration of a compound in the solution that surrounded the cell. If a root hair cell, for example, was placed in a 7 percent sucrose solution the cell "will show either no plasmolysis or a very weak plasmolysis" (Overton, 1899 in Branton and Park, 1968b: 45). However, when the sucrose concentration was increased to 7.5 percent, "a uniform plasmolysis takes place within ten seconds" (Overton, 1899 in Branton and Park, 1968b: 45). Moreover, the state was reversible. A root hair immersed in a solution of distilled water showed "a spontaneous disappearance of the plasmolyzed condition" (Overton, 1899 in Branton and Park, 1968b: 46). Acting on the assumption that a selectively permeable barrier was responsible for the observed effects---"for the last three years I have...suspected that the peculiar osmotic properties of the living protoplasts are due to a *selective solubility*"---Overton reasoned that if a compound could rapidly traverse the membrane, the concentration equilibrium would quickly be established and plasmolysis would be minimal (Overton in Branton and Park, 1968b: 47; emphasis in original). This assumption marked a turning point in Overton's experiments---and membrane studies in general---because it ultimately led to a model of cell membrane molecular structure.

Overton experimented with a wide variety of compounds and discovered a pattern in the results. Compounds soluble in fatty oils entered the cell very quickly while compounds soluble in water but "insoluble or very slightly soluble" in fatty oil either penetrated the cell very slowly or did not penetrate it at all. This finding was strengthened by a second series of experiments "following this effect" in which Overton substituted molecular groups in a number of experimental compounds as a hypothetical means of increasing their solubility in fatty oil (and thus the rate of penetration). In a typical experiment, the urea molecule, which normally enters the cell very slowly, was found to penetrate very quickly if a methyl or ethyl group was substituted for one of the hydrogen atoms. If two hydrogen atoms were replaced by two methyl or ethyl groups, the penetration time was even faster. If three groups were substituted, penetration occurred "almost immediately". Overton concluded that it was "very probable that the general osmotic properties of the cell are due to the end layers of the protoplast, which are impregnated by a substance whose dissolving properties for various compounds may well match those of a fatty oil" (Overton in Branton and Park, 1968b: 49). Overton admitted uncertainty as to the precise composition of what he termed a "precipitation membrane", but speculated presciently that it was "not a normal fatty oil". He suggested that the membrane was somehow composed of cholesterol and fatty oils as "the impregnating substance" (Overton in Branton and Park, 1968b: 49). The concept of a lipid membrane as a selective cellular barrier was introduced to biology.

Overton's inability to describe membrane structure in anything other than the most general terms was due to the limits of instrumentation. At the time of his research, no devices existed either for observing the membrane *in situ* or in an artificial preparation. The former limitation would not be overcome until the development of electron microscopy in the 1930s and 1940s and subsequent application (by Albert Claude, Keith Porter and George Palade) to biological specimens in the 1940s and 1950s (Branton and Park, 1968a; Wallace, 1996; Cooper, 1997). Research on the latter problem, however, was already underway in Germany eight years before Overton's presentation before the Science Foundation in Zurich. In 1891, Agnes Pockels, a highly talented investigator with no formal training in physics or chemistry---she conducted experiments in her kitchen sink using bowls, strings, and buttons---had developed a rectangular trough for measuring surface tension in liquids (Tanford, 1989). Due to the support of Lord Raleigh, a Cambridge physicist who had conducted experiments measuring the expansion of a known volume of oil on the surface of water, Pockel's results were published in *Nature*. She followed this initial effort with a series of publications describing the novel device in detail. These eventually caught the attention of the chemist Irving Langmuir who, with his assistant Katherine Blodgett, was conducting research for General Electric on molecular monolayers. Langmuir and Blodgett improved Pockel's apparatus, known today (somewhat unfairly) as a Langmuir-Blodgett (LB) trough, and then performed a decisive experiment involving a film of lipids (Langmuir, 1917; Kanicky and Shah, n.d.). Using the improved trough, Langmuir demonstrated that lipids with various hydrocarbon-chain lengths would produce a film of the same area on the surface of water. This could only be possible if the molecules were vertical. The most chemically plausible orientation would be for the polar head groups to be in contact with the water and the nonpolar hydrocarbon chains pointing upward. This conclusion was a critical piece of evidence for the amphiphilic structure of lipids, a concept which played a major role in the understanding of membrane structure.

The new understanding came from the Dutch biologists Edwin Gorter and F. Grendel, who conducted a decisive experiment with erythrocytes. Using acetone and other solvents, they extracted 1 cc of lipids from red blood cells taken from dog, sheep, rabbit, guinea pig, dog, and human. Gorter and Grendel noted the significant difference in the sizes of the erythrocytes of these species, but argued that the smaller cells (from goat and sheep) were present in greater quantities than the larger cells (from dog and rabbit) thereby producing an approximately equal total surface area (Gorter and Grendel in Branton and Park, 1968). (In fact, this assumption caused them to underestimate the total surface. But, as was subsequently recognized, their error was cancelled out by their faulty lipid extraction techniques, thus leading to a result which was essentially correct [Sadava, 1993]). Having extracted the lipids, Gorter and Grendel then applied them to the surface of water in a modified version of Langmuir's trough. Based on Langmuir's findings, the researchers knew that the lipids would "form precisely a monomolecular film" (Gorter and Grendel in Branton and Park, 1968: 54). The measured area, however, was approximately twice the total surface area of the erythrocytes. This could only be possible if the cell membrane were a bilayer: "One obtains [he wrote] a quantity of lipoids that is exactly sufficient to cover the total surface of the chromocytes in a layer that is two molecules thick…We therefore suppose that every chromocyte is surrounded by a layer of lipoids, of which the polar groups are directed to the inside and to the outside" (Gorter and Grendel in Branton and Park, 1968: 53). The supposition was later verified directly when they squeezed the edges of the lipid film together using wooden floats. The film was observed to double up on itself and form a bilayer of the type they had hypothesized.

2.4. The Davson-Danielli and Fluid Mosaic Models

Gorter and Grendel's success in identifying the bilayer structure of the membrane immediately triggered an intensive series of experiments and conflicting interpretations regarding membrane functional properties. Of all the models proposed in the 1930s, the most influential was surely that of Hugh Davson and James F. Danielli (1935). They argued that the membrane was essentially a sandwich-like structure with protein surfaces and a lipid interior. To some extent, this viewpoint derived from Danielli's earlier research at Princeton with E. Newton Harvey (1933-1935) in which it was observed that an oil drop inside a mackerel egg had reduced surface tension. Danielli and Harvey had suggested that this effect was due to protein adsorption to the surface of the oil drop. Drawing upon this earlier model, Danielli and Davson (1935) maintained that a protein coat would explain the fact that a pure lipid membrane absorbed water more slowly than a membrane associated with proteins. Since most proteins are water-absorbent, it appeared plausible that interior and exterior protein coats expedited the transmembrane movement of water. Their viewpoint was (initially) reinforced in the 1940s and 1950s by findings based on the newly-developed techniques of electron microscopy (EM). The optical resolution in early EM had been reduced to approximately 1 nanometer, far exceeding the 550-nanometer resolution of the light microscope. When the new protocols were applied to the cell membrane, a trilaminar membrane structure of dark, light, and dark bands was revealed. The dark bands (often described as "railroad tracks") were immediately believed to be the protein coats, while the light band was construed as the lipid interior. The Davson-Danielli (DD) model now seemed to have strong experimental support.

Yet even while Davson and Danielli were formulating their model, evidence incompatible with their viewpoint was accumulating. The basic problem was the known rigidity of protein lattices, a feature difficult to reconcile with molecular movements across the membrane. J. David Robertson (1957) had attempted to solve this problem by proposing protein pores in his "unit membrane model", but this modification had left intact the putative protein lattice, and was therefore insufficient to accommodate problematic new findings. Microsurgical techniques made it possible to observe a membrane when its external surface is gently pushed with a probe. It was noted that the membrane would respond to the probe elastically, bending inward somewhat like a balloon, and then relaxing, when the probe was released, back to its original shape. Moreover, if the probe actually penetrated the membrane, the membrane would conform around the probe and then reseal when it was removed. These findings strongly suggested a fluid-like membrane structure and not a rigid protein lattice. The latter, it was argued, would be less elastic and would probably tear, due to breaking of bonds, when penetrated by the probe. Further difficulties became apparent as the molecular composition of the membrane was better understood. Freeze-fracture techniques, in which membrane samples are frozen in liquid nitrogen and fractured with a knife blade, indicated the presence of proteins in the middle of the membrane. Thus the proteins could not be a coat, contrary to the DD model. Subsequent, more detailed, analyses of the proteins revealed that they were amphiphiles, with hydrophobic and hydrophilic segments. This feature would have resulted in thermodynamic instability if the proteins were in contact with extracellular or cytosolic water. The only obvious way in which the membrane structure could be at minimum local potential energy would be for the protein hydrophobic segments to be adjacent to the lipid hydrocarbons, and the protein hydrophilic segments to extend beyond the membrane.

The decisive blow to the DD model was an elegant experiment conducted by L.D. Frye and M. Edidin (1970) with fluorescent-labeled antibodies. In their study, two different cell types, each containing a differently labeled antibody (green or red) were fused and observed following a 40-minute interval. The hybrid cells displayed a green-and-red speckled "mosaic" indicating membrane fluidity. This experiment provided the strongest empirical support for what has since become known as the "Fluid Mosaic Model" (FMM). Two years after the Frye-Edidin experiment, Jonathan Singer and Garth Nicolson proposed FMM, which has continued (with the important modifications discussed below) as the dominant model to the present time (Singer, 1972; Singer and Nicolson, 1972). FMM's defining feature is a membrane construed as a 2-dimensional fluid medium in which the constituent lipids form random associations. Their model also incorporated the amphiphilic structure of proteins, the discovery of which had contributed to the downfall of the DD model. Singer and Nicolson distinguished two classes of membrane-associated proteins: peripheral and integral. The former were operationally defined as proteins that dissociate from the membrane (without disrupting the bilayer) following a treatment with a polar reagent such as a solution with high salt concentration (Singer and Nicolson, 1972; Cooper, 1997). The latter, by contrast, could only be dissociated by amphipathic molecules (e.g., detergents) which would displace the membrane lipids and bind to the hydrophobic portions of the proteins, thereby forming soluble complexes. The functional implication of FMM was that proteins were largely responsible for chemical activities at the cellular surface. The stochastic interactions in the membrane---Singer and Nicolson had stated that lipids form "random associations"---were difficult to reconcile with the highly organized interactions involved in (for example) transmitter-receptor binding, enzyme-substrate activity, and other molecular systems. As a result, the role of the membrane was limited to that of a semi-permeable barrier separating aqueous compartments and serving as an anchoring site for proteins. Investigations of the last two decades, however, suggest that this view may be in need of revision. The modifications of FMM will be examined in the final section.

2.5. The Fluid Mosaic Model Questioned: Chistyakov and Kinnunen

From the earliest days of its formulation, possible inadequacies were noted in FMM.

In 1967---five years prior to Singer and Nicolson's FMM---the crystallographer I.G. Chistyakov had proposed in a review article in *Soviet Physics* that any given region within the cell membrane appeared to alternate between molecular disorder and order; i.e., that the membrane had the bulk property of phase transitions between liquid and crystalline states (Chistyakov, 1967). Possibly because the article had appeared in a journal somewhat less widely read than many of its European or American counterparts, Chistyakov's observations for a time passed largely unnoticed (Wallace, 1996). Throughout the 1970s and 1980s, however, a large number of experiments (summarized briefly below) conducted with natural and artificial membranes lent support to Chistyakov's viewpoint. These data became the basis for two pioneering articles in which Paavo Kinnunen of the University of Helsinki argued that the transiently organized states in cell membranes: 1) were the evolutionary result of the combinatorial complexity associated with a cellular boundary-maintenance system; 2) played a significant role in many cellular functions; 3) regulated ion-channel dynamics in neuron signal processing (Kinnunen, 1991; Kinnunen and Virtanen, 1986). It may be useful for the historical appreciation of more recent membrane models (e.g., "raft" or "microdomain" theories) to examine each of these claims in detail.

Kinnunen (1991) was the first to propose that the membrane likely possessed sophisticated regulatory features consistent with a semi-permeable boundary. The membrane would have to be capable of a set of adaptive responses to external and intracellular changes in temperature, pH, osmolarity, as well as binding with metabolites, hormones, and other agonists. Because many different combinations of these extracellular and cytosolic inputs are possible, there would have been strong selection pressure on early membranes to develop local variations in molecular architecture. Kinnunen argued, presciently, that membranes were probably divided into "compositionally and functionally different domains" separated by mechanical barriers such as the cytoskeleton. (Intriguingly, he further speculated that functional incompatibility between early membrane domains required the emergence of the separate, largely autonomous, membranes of the endoplasmic reticulum and mitochondrion). Consistent with these claims was the great diversity of lipid species in eukaryotic cells in contrast with those of prokaryotes. Thus *E. coli* membranes contain approximately ten different phospholipids in contrast with approximately 100 in eukaryotic cells. If minor phospholipids and metabolic intermediates are taken into account, the numbers for *E. coli* and eukaryote are 100 and 1000, respectively. "Accordingly," Kinnunen argued, "we may hypothesize that crucial in the development of eukaryotes was the more efficient exploitation of the properties of lipids in the functions of cellular membranes, achieved through the evolution of the diversity in lipid species."

Consistent with his evolutionary viewpoint, Kinnunen proposed that a wide variety of cellular functions appeared, to a significant extent, to be membrane-regulated. Although no single specific mechanism of membrane regulation was identified, strong emphasis was placed on the possible interaction of integral membrane proteins with neighboring "boundary" lipids. Experimental studies indicated that the proteins GAP (*ras* GTPase activating protein) and α-actinin were sensitive to the acyl chain composition of the boundary lipids. Similarly, enzyme activities as well as ligand specificities and affinities [e.g., the function of the vitronectin receptor (VnR) which plays an important role in regulating tumor cell invasion and dissemination (Baroni et al., 2003)] could be altered by changes in boundary-lipid composition. In view of these interactions (and similar modulating effects demonstrated for a large number of peptides) Kinnunen concluded that the randomness posited in FMM was probably not present: i.e., even when a membrane region was in an unperturbed state, spatio-temporal "domains" would exist "corresponding to the free energy minimum of the system". This was one of the earliest claims---perhaps, in fact, *the* earliest---that the membrane was not a "sea of lipids" but a ordered structure of microdomains separated by regions of interfacial fluidity. The latter view, now supported by a growing body of experimental evidence (discussed below), may have important implications for membrane control of neuron signaling.

Kinnunen's final contribution (in association with J.A. Virtanen and historically five years earlier) was a model of membrane regulation of the Hodgkin-Huxley (HH) action potential (Kinnunen and Virtanen, 1986). In order to appreciate the novelty of their viewpoint, it may be useful to review the basic features of the HH model. As demonstrated by Alan Hodgkin and Andrew Huxley in a series of landmark experiments conducted on the squid axon in 1952, a "large, brief, invariant signal", which they termed an action potential, propagating along an axon without decrement is the physical basis of information in the neuron and, by extension, in the nervous system (Sargent, 1992; Hodgkin and Huxley, 1952a; 1952b;1952c;1952d). The signal is a change in membrane potential (voltage inside the neuron in relation to voltage outside) from a resting potential of ~ -65mV to ~ +55 mV (at the peak of the action potential) and back to –65mV in less than 1 msec. The change in potential results from the opening of voltage-gated Na^+ channels and the inward movement of ions through each channel at the rate of 0.6-12 X 10^7 ions/sec (corresponding to a current of 1-20 pA). At the Na^+ equilibrium potential of ~ +50 mV, the Na^+ channels are inactivated and voltage-gated K^+ channels open. Potassium ions then flow out of the neuron, leading to brief hyperpolarization at ~ -75 mV followed by restoration of the resting potential at ~ -65mV. The HH model (explained in more detail in any neuroscience textbook) has been verified by thousands of experiments. It remains the conceptual keystone of modern molecular neurobiology.

Although Kinnunen and Virtanen (KV) did not question the HH model's usefulness for understanding neural systems at higher orders of magnitude (e.g., the network or tissue levels), they did suggest that within the neuron there were smaller information-processing mechanisms. Their argument was primarily based on charge-transfer (CT) theory applied to unsaturated membrane lipids. They noted that a typical feature of membrane lipid hydrocarbon chains is an unsaturated bond (C=C) between carbon atoms 9 and 10. Each unsaturated region---the ethene portion of the lipid hydrocarbon chain---is a p orbital which forms a σ bond with the ethene of the adjacent lipid. There is a high degree of orbital overlap, consistent with high conductivity of electrons---an important feature of the model. In this manner, lipid ethenes form a linear array parallel to the plane of the membrane. CT theory predicts that if one extra or an extra pair of electrons is introduced at one end of the aligned ethenes, the electrons will propagate through the array. In the KV model, the specific reactions that introduce electrons into the array are not described in detail. Rather it is noted that the inward movement of Na^+ ions causes the dissociation of a cholesterol-phospholipid complex. The dissociation, in turn, releases a spin-correlated electron pair (or quasi-particle) into the linear array.

Somewhat disappointingly, the KV model ends at that point. No specific means is identified by which a propagating quasi-particle might regulate ion channel dynamics. Equally important, there is no discussion of a reset mechanism: i.e., no process is outlined by which the correlated electrons would either return to their source or exit from the array to be recycled. The author attempted to address the first shortcoming (at least to a limited extent) by proposing that the field propagated orthogonally to the direction of quasiparticle movement (or to bilayer normal) would briefly promote outer electrons of the linear ethene array to metastable (Rydberg) states (Wallace 1996a, 1996b). In this speculative proposal, "configurations" of charge in the promoted ethene orbitals would interact electrostatically with one another and with charge centers in an unfolded ion-channel protein. It was then claimed, but not clearly shown, that these electrostatic interactions could regulate the mean open time of the channel. Moreover, this variation of KV did not solve a more fundamental problem. As later investigations (discussed in detail below) suggested, the onset of permeant ion movement at the onset of the HH action potential most likely generates the formation of unsaturated *and saturated* membrane microdomains (Groves et al., 1997; Radhakrishnan and McConnell, 2000). Accordingly, the picture of a "pathway" of σ-bonded p orbitals of adjacent ethenes along which a quasiparticle is propagated is probably not correct.

2.6. Ole Mouritsen and Myer Bloom: The Hydrophobic Mismatch (HM) Model

At the same time that the KV model was being developed, Ole Mouritsen and Myer Bloom were arriving at their own novel concept of phase behavior in membranes (Mouritsen and Bloom, 1984; 1993). Although the model is currently controversial, it remains one of the most persuasive explanations for membrane lipid-protein interaction (Killian, 1998; Dumas et al., 1999). Its fundamental structural feature is the hydrophobic length of an integral protein and the hydrophobic thickness of the lipid bilayer core. If the membrane is quiescent (e.g., unperturbed by ligand binding or an applied electric field) there is a high degree of matching between the core of the bilayer (lipid hydrocarbon chains) and the membrane-spanning domain of the protein, both of which are hydrophobic. However, if the hydrophobic region of the integral protein or that of the membrane lipids is exposed to water due to electrical or chemical perturbation, membrane lipids reorganize to reduce the free energy of the system. Lipid species with hydrophobic lengths most closely approximating that of the protein become more abundant in the protein's vicinity, a process known as "interface enrichment". The result of the reorganization is a transient ordered state, i.e., a microdomain. When the perturbation is removed, the membrane lipids return to the disordered state. The model is generally known as the Hydrophobic Mismatch (HM) Model, although Mouritsen and Bloom originally called it "the mattress model"---a metaphorical reference to the protein-induced fluctuating thicknesses in the membrane.

The developed form of HM, published in 1993, was primarily supported by the predictions of a computer model compared to the findings of a (then) small number of studies of hydrophobic mismatch in reconstituted lipid vesicles (Mouritsen and Bloom, 1993). The computer simulation was strongly indebted to D.A.Pink's 10-state model which construed the membrane as a statistical-mechanical lattice and incorporated ten conformational states of lipid diacyl chains (Pink and Chapman, 1979). Mouritsen and Bloom extended those states to include lipid-protein interactions, and assumed that the membrane-spanning portion of the protein was a "stiff, rod-like, and hydrophobically smooth object with no appreciable internal flexibility." Experimental data regarding pure bilayer properties as well as the 3D geometry of proteins (to the extent that the structures were then known) was fed into the model. The results were then compared with experimental determinations (obtained by H. Möhwald and colleagues) of hydrophobic mismatch in photosynthetic reaction center proteins and light-harvesting chlorophyll proteins (antenna proteins) reconstituted into lipid bilayer vesicles (Peschke et al., 1987). The result was a satisfactory fit between theory and experiment, an outcome particularly gratifying given the researchers' admitted uncertainties regarding protein 3D geometries and other measurements in the experimental data set.

Additional experimental support for the HM model was not long in coming. Three years after the above study, Ole Mouritsen and Paavo Kinnunen (the latter the co-author of the KV model) summarized a substantial body of evidence indicating that membrane enzymes and channels "display optimum activity at a certain bilayer thickness, which supposedly is the thickness providing the better hydrophobic match" (Mouritsen and Kinnunen, 1996). Good examples of this optimizing effect were cytochrome c oxidase, (Na^{+} -K^{+})-ATPase, and Ca^{2+} -ATPase. In addition, pyrene-labelled phospholipids in membranes reconstituted with the *E. coli* integral protein lactose permease displayed a lateral concentration gradient consistent with the HM model [See also the later study by Lehtonen and Kinnunen (1997)]. But perhaps the most forceful examples for lipid mediation via HM of integral protein function were gramicidin A and rhodopsin. The former is a polypeptide of 15 hydrophobic amino acids, in alternating D- and L- forms in a helical structure around an aqueous pore, and with N and C terminals blocked such that the molecule is nonpolar and soluble in the membrane. The lifetime of this channel can be decreased by increasing the hydrophobic mismatch, a finding consistent with a later study by Niloufar Mobashery and colleagues indicating that the gramicidin conformational preference is sensitive to membrane thickness (Mobashery et al., 1997). In similar fashion, the transition between the metarhodopsin I and II conformational states of rhodopsin could be experimentally manipulated by modifying hydrophobic mismatch via lipids with non-lamellar propensity.

2.7. Hydrophobic Mismatch and Neural Evolution: Wallace and Price

The growing evidence that hydrophobic mismatch may be a fundamental means by which membranes regulate integral protein function suggested a possible application to ion-channel activity in neurons. In 1999, Harry Price and the author presented the above data set and proposed that hydrophobic mismatch would be routinely generated in the neuron by ion-channel protein conformational change from the energetically stable α-helix conformation to the random-coil state associated with ligand or voltage gating (Wallace and Price, 1999; Wallace, 1999). Membrane lipid reorganization would occur in the channel protein's vicinity to reduce the degree of mismatch. The resulting lipid-protein ensemble would be a momentarily poised system in which the relaxation of the ion-channel protein to the α-helix conformation would be momentarily opposed by condensed membrane lipids. Subsequent protein relaxation would re-establish the original resting state in which the ion channel is closed. The membrane would then return to the fluid state. Because protein folding does not begin (in this model) until membrane reorganization is complete, the time required for lipid self-assembly would directly regulate the mean open time of the channel. The collective effect of many such systems operating in the neuron would be the propagation of the HH signal. The conclusion was thus consistent with Alwyn Scott's earlier suggestion in a popular study of brain systems that the neuron may not be a switch but a structure "composed of many smaller switches" (Scott, 1995).

In the second part of their argument, Wallace and Price examined the possible role of HM-induced neural membrane molecular order in brain evolution. They reasoned that if the targeting and molecular composition of neural-membrane microdomains was to some extent genetically regulated, and if the microdomains played a significant role in neuron signaling (and, hence, in an animal's survival), then the microdomains should evolve like any other aspect of phenotype. This second part of the argument thus focused on the secretory pathway. This is the intricate cellular system involving (principally) the endoplasmic reticulum and Golgi by which small proteins or polypeptides are post-translationally bound to vesicles which are then cytoskeletally targeted to the membrane. The Price-Wallace model reflected the understanding of this process in the 1990s. During that decade a series of experiments conducted with Madin-Darby canine kidney (MDCK) cells, hepatocytes, and intestinal epithelial cells indicated that polypeptides were sorted in the Golgi complex, inserted into lipid vesicles, and routed to membrane targets (Geiger et al., 1992; Nelson et al., 1992). The latter process might be direct or indirect. In the case of MDCK cells, vesicles are routed directly to the membrane. Hepatocytes and intestinal epithelial cells, by contrast, are initially delivered to the membrane and then re-routed to the target region by endocytosis. Together these data suggested a fairly close regulatory interplay between gene expression, enzyme activity, and organelle function. However, there were comparatively few studies of the secretory process in neurons. Investigations by Simons et al. (1992) had demonstrated that lipid vesicles transport small proteins from the Golgi exit to either the somatodendritic region or the axonal region of nerve cells. But apart from this pioneering study, little work had been done.

The final part of Wallace and Price's argument was a proposed outline of a protocol for membrane microdomain studies in evolutionary neurobiology. The research would be conceptually based on the basic Darwinian principle that sustained natural selection pressure on a phenotypic trait is an optimizing process. The principle of evolutionary optimization may be succinctly stated as follows: successive elimination of less adapted phenotypes in an initially more varied population over many generations results in highly adapted phenotypes in a much less varied population. In a way, the varied phenotypes in any given generation resemble the varied outputs in artificial learning systems (Boltzmann machines). Continuing the analogy, elimination of the less fit phenotypes is similar to error reduction which is "fed back" to the population's gene pool in the form of changed genetic composition. Although the basic concept of biological optimization dates back (at least) to Charles Darwin, it has been recently popularized by biologist Richard Dawkins (1986; 1996). In addition, the similarity to learning systems has caught the attention of engineers. A formal comparison of evolution to error minimization in artificial-intelligence systems was developed by Wirt Atmar (1994). Wallace and Price utilized a modified version of Atmar's formalism in which a biological system may be described by the alphabet $\{ Q, I, Z, \delta, \omega \}$ where Q_i is the system's internal states, I_i is environmental inputs, Z_i is output values, δ is a next-state mapping function such that $\delta: I_1 X Q_1 \rightarrow Q_2$. Error E may then be expressed as the sum of accumulating environmental mispredictions

$$E = \sum | i_{t+1} - z_t | \qquad (1)$$

where

$$\lim(t \rightarrow \infty) \sum | i_{t+1} - z_t | = 0 \qquad (2)$$

is adaptive global optimization. Thousands of such systems have been described in evolutionary biology. In humans, examples would include retrieval latencies in memory, probability of recall for most highly needed information, and predicting the features of novel objects (Anderson, 1991a, b).

The alphabet was then used to describe possible comparative experiments conducted on membrane micelles to investigate optimization in solving complex problems (Garey and Johnson, 1979). The studies would be conceptually based on an assumption made by Kinnunen (1991) that any membrane microdomain is highly adapted to process combinatorially complex external and cytosolic inputs. Micelle models constructed from the most typical lipid ratios in tissue samples extracted from the same brain micro-region in different animal species would be constituted with multiple receptors and ligand-gated. Analytic techniques such as scanning fluorescence correlation microscopy or time-resolved x-ray diffraction would provide detailed read-outs regarding lipid-protein dynamics (Edidin, 1989; Petersen et al., 1993). It was predicted that micelles would differ in their responses to escalating input complexity (i.e., ligand-gating of an increasing number of receptors). The more optimized membrane systems would continue to display distinctive fluoroscopic signatures for successive sets of gated receptors when the less optimized systems were displaying a plateau response. The latter signal would describe a situation of asymptotic searching for a local energy minimum. In terms of the Atmar-based formalism, given a distinctive signature S of successive membrane states S: Q_1 , Q_2 ,. . . Q_N associated with escalating complexity of input I: $[I_1]$, $[I_{1,2}]$, $[I_{1,2,3}]$, . . .$[I_N]$ (where I_i designates total number of gated channels) the more computationally sophisticated membrane systems would be characterized by a higher number of dedicated signatures; i.e., by more $S \rightarrow Z$ read-outs before arriving at the plateau response.

The Wallace-Price model (schematically summarized here) was an attempt to operationalize the idea that neural membrane microdomains may be evolutionarily optimized regulatory systems. It thus proposed a molecular complement to the classical approach which currently dominates computational neurobiology and brain evolutionary studies. This objective was both justified and plausible, but the data upon which it was based appears now to have serious shortcomings. Viewed in the light of more recent research, the model's fundamental claim---that the neural membrane is a computational system---may well have been right for the wrong reasons. Too much significance may have been attributed to the hydrophobic mismatch mechanism. Although HM may well be crucial for a number of cellular functions including (importantly) protein sorting in the Golgi system, it may not be the only nor even the most significant physical basis for membrane self-organization---particularly in the neuron. The evidence against the HM model will be examined below.

2.8. Limitations of the Hydrophobic Mismatch (HM) Model

Much of the fascination (and no small part of the frustration) of membrane biophysics is due to the fact that the lipid bilayer and associated integral proteins follow several different strategies to deal with hydrophobic mismatch (Killian, 1998). For example, protein aggregation in response to this perturbation is predictable on theoretical grounds and has been confirmed experimentally. Bacteriorhodopsin reconstituted in saturated and unsaturated phospholipid membranes responded by aggregation to HM in which the bilayer thicknesses were 4 Å thicker and 10 Å thinner than the hydrophobic length of the protein (Lewis and Engelman, 1983). Smaller mismatches were tolerated, the researchers noting in this regard that "the lipid bilayer can evidently sustain large local distortions with a small change in free energy." Similarly, skeletal sarcoplasmic reticulum (Ca^{2+})-ATPase reconstituted in phosphatidylcholines with chain lengths varying from 14-24 carbons demonstrated maximal activity with an 18-carbon lipid chain length, responding to greater or lesser lengths with protein aggregation (Cornea and Thomas, 1994).

In like fashion, the protein-membrane molecular ensemble can respond to HM by a tilt of the protein backbone with regard to bilayer normal. This response has been observed for integral membrane proteins with multiple membrane-spanning segments. In a recent experiment, Le Coutre et al. (1997) identified an average tilt angle of 33° for the helices of lactose permease from *Escherichia coli* reconstituted in a lipid bilayer in which the lipid-to-protein ratio ≈800:1. With increasing protein content (and paralleling decrease in bilayer thickness) an increase in helix tilt occurred. Moreover, the degree of tilt appeared to be correlated with protein function. Ion channels may also respond by chain tilt to hydrophobic mismatch. This possibility was strongly indicated by the fluorescence-emission studies conducted by Williamson et al. (2002) to examine the response of the *Streptomyces lividans* KcsA potassium channel to changes in bilayer thickness (Williamson et al., 2002). The investigators found that the KcsA tryptophan residues "form bands on the two sides of the membrane with the rings of the Trp residues being almost parallel to the surface of the membrane." This observation is consistent with other studies indicating that Trp residues of multiple membrane-spanning proteins tend to be located at the lipid-water interface (Killian, 1998). As the investigators manipulated the membrane thickness (from C14 to C18) the Trp residues moved away from the center of the bilayer as the tilt of the helices decreased to match the thicker membrane. The thermodynamics of the tilt response is presently not well understood. However, De Planque et al. (2001) speculate that the "disruption of membrane packing caused by the tilt of one helix could be compensated for by tilt in the opposite direction of a second helix." In any event, it is becoming evident that lipid reorganization and "interface enrichment" as proposed in the HM model appear to occur only to a limited extent in the case of complex proteins with multiple membrane-spanning segments, if indeed they occur at all.

2.9. The Ordered Membrane: Fluorescence Studies

At the end of the 1990s and the beginning of the present century, a series of fluorescence experiments triggered a significant change in thinking regarding membrane organization. Since the 1970s, Singer and Nicolson's Fluid Mosaic Model (FMM) had dominated thought on the subject (Singer and Nicolson, 1972). As we have seen, the fundamental feature of FMM was a presumed resting state of molecular disorganization analogous to a 2D liquid. Put differently, it was assumed that the liquid state corresponded to a local potential energy minimum. Molecular organization was thus transient (10^{-4} s), resulting from a perturbation such as conformational change in a membrane protein or from an applied electric field. Membrane order was also spatially limited to microdomains of ~1-30 nm lateral length isolated in a fluid medium (Kinnunen, 1991; Lehtonen et al., 1996; Dumas et al., 1997; Lehtonen et al., 1997). As a result of fluorescence studies, this model has been precisely inverted. As Hao et al. (2001) have recently expressed it, "instead of having a few rafts of ordered lipids in a sea of fluid lipids, it might be more accurate to envision *a membrane that is mostly rafts* with small regions of fluid lipids intercalated among mostly ordered lipid domains" (Emphasis mine). The effect of chemical or electrical perturbation is thus not the generation of microdomains, but rather their coalescence into domains ranging from hundreds of nanometers to microns in length and stable for tens of minutes. As will be discussed in the following chapter, this challenge to FMM may prove to be highly significant for the regulation of neuron signaling.

The evidence supporting the claim of Hao et al. (2001) that the membrane "is mostly rafts" is highly convergent. A large number of fluorescence studies indicate that the presence of cholesterol and hydrocarbon-chain saturation promotes "highly ordered and densely packed [molecular] assemblies" (Li et al., 2001; Mitchell and Litman, 1998; Wang et al., 2000, 2001; Samsonov et al., 2001). The tetracyclic hydrocarbon ring system of cholesterol plays a significant role in this packing process (Brown, 1998). The rings, which are in a compact and planar conformation, are adjacent to the hydrocarbon segments of sphingolipids. In that position, the ring system diminishes the isomerization and reduces the "kinks" of the adjacent sphngolipid hydrocarbon chains. The net result of the latter changes is increased molecular compressibility. Mitchell and Litman propose that the preferential interaction of cholesterol with saturated-chain lipids drives them to the center of the microdomain (Mitchell and Litman, 1998). The result is a liquid-crystal phase consisting of rapidly forming and dispersing saturated-lipid/cholesterol microdomains connected by an interfacial region consisting primarily of polyunsaturated lipids. Several investigators have noted that the segregation of the microdomains is cholesterol-dependent (Li et al., 2001; Brown, 1998; Mitchell and Litman, 1998; Hao et al., 2001; Samsonov et al., 2001). Hao et al. have proposed a more generalized interpretation where "small changes in physicochemical variables such as cholesterol content, changes in lipid head group chemistry, or protein interactions might induce growth or coalescence of certain types of lipid domains. Such a change could in turn initiate membrane-associated physiological phenomena such as signal transduction."

The final major contribution to the revised view was the demonstration by Samsonov et al. (2001) that microdomains are co-localized between membrane monolayers. In the initial stage of their investigation they noted that dark circular regions formed in planar bilayer membranes when sphingomyelin and cholesterol were present, and did not form when either was absent. To determine whether or not the dark regions were rafts, the researchers attached a fluorescent probe to the ganglioside GM_1 which preferentially partitions into cholesterol-sphingolipid microdomains. When the probe was present, the circular regions were bright, confirming that the previously observed dark regions were cholesterol-sphingomyelin rafts. In the final stage of their study, they observed that the dark circular regions (microdomains) did not overlap. This was direct visual evidence that microdomains "within the two monolayers interact to form a coherent unit". This finding is consistent with the earlier emphasis placed by Kinnunen (1991) on localized membrane regions as mediators of extracellular and cytosolic inputs.

This model of an ordered, dynamic, laterally compartmentalized membrane, arrived at by a tortuous route beginning early in the nineteenth century, may ultimately prove to be significant---as Hao et al. (2001) have noted---for the understanding of a wide variety of cellular mechanisms. Zajchowski and Robbins (2002) concur, summarizing a large body of evidence for significant roles of membrane microdomains in endocytosis, cholesterol transport, calcium homeostasis, protein sorting, growth factor signaling pathways, and the internalization of toxins, bacteria, and viruses. In the case of the neuron, the new viewpoint may possibly become the basis for a changed understanding of ion-channel dynamics, transmitter exocytosis, and other features of electrochemical signaling (Tsui-Perchala et al., 2002). This possibility---and its implications for the understanding of neural disease---will be examined at length in the remainder of this book.

Chapter III

Membrane Microdomain Regulation of Neuron Signaling

This chapter presents a model of membrane microdomain regulation of neuron signaling. It is based on laboratory investigations of natural and artificial membranes as well as computer-simulation studies conducted by Harry Price in collaboration with the author (Price and Wallace, 2001; 2003). The model emphasizes the role of the membrane in the regulation of ion-channel dynamics. The fundamental feature of the model is field-induced reorganization of the membrane lipid bilayer. It is proposed that transmembrane ion movement through an "open" ligand- or voltage-gated channel generates an electric field which propagates in the plane of the membrane. The membrane to which the field is applied is already in an ordered state consisting of sphingolipid/cholesterol microdomains separated by a fluid interface of unsaturated lipids. In response to the applied field, microdomains coalesce and fluid, unsaturated lipids become densely packed. Ethenes in cholesterol/sphingolipid microdomains seek an energetically favored alignment in the plane of the bilayer. In this aligned or "stacked" conformation, the ethenes become highly polarized. Electrostatic interactions between the aligned-ethene dipoles and polar amino acids in the unfolded ("open") ion-channel protein regulate the rate of protein folding (ion-channel closing). In this manner the membrane controls the duration of the ion-channel open state. The collective effect of this lipid-protein interactional system operating at multiple ion-channel sites is the regulation of the HH action potential. The model will be presented in detail below.

3.1. The Resting Membrane is in a Liquid-Ordered State

As we have seen in the previous chapter, a biological membrane in an unperturbed state is a liquid-ordered structure. Microdomains of cholesterol and sphingolipids with estimated lateral lengths of ~10-300 Å and associative lifetimes of 10^{-4} s are separated by fluid interfaces comprised of polyunsaturated lipids (Mouritsen and Jørgensen, 1992; Jørgensen and Mouritsen, 1995; Marsh, 1995; Simons and Ikonen, 1997). The sphingolipids and cholesterol of the microdomains are in a densely packed aggregation. The dense packing is due to favorable interaction between the steroid ring of cholesterol and the saturated hydrocarbon chains of the sphingolipids as well as hydrogen bonding between the 3-OH group of cholesterol and the amide bond of the sphingolipids (Brown, 1998). Within each sphingolipid are two hydrocarbon-chain ethylenic bonds. These are proposed to play a significant role in the regulation of neuron signaling. One bond is located close to the extracellular and cytosolic interfaces and attaches an invariant 15-carbon fatty acid chain to the sphingolipid head group. The other bond is usually located between C9 and C10 in a variable fatty acid chain ranging in length from 10-22 carbons (Kinnunen and Virtanen, 1986; Zuckermann, 1993). In the resting state, microdomains are constantly in flux, rapidly forming and dispersing. This liquid-ordered state changes pronouncedly during ion-channel gating.

3.2. Neural-Membrane Microdomains Respond to an Electric Field

The ion channel of a neuron may be chemically or electrically gated (Hille, 2001; Nestler et al., 2001). In the former case, an inhibitory or excitatory transmitter molecule binds with an ion-channel receptor site, thereby changing the conformation of the channel protein from the closed (α-helix) to the open (random coil) state. In the latter case, an ion-channel voltage sensor (a transmembrane domain of positively-charged arginine or lysine residues) moves in response to an applied electric field, triggering conformational change to the open state. In either event, there is an immediate movement of ions through the pore of the open channel. The current flowing through the channel is normally in the 1-20 pA range, corresponding to a movement of ~.6-12 X 10^7 ions/s (Levitan and Kaczmarek, 2002). The transmembrane ion movement generates a field in the bilayer of 100 kV/cm or 10 V/μm. The applied field may have dramatic effects on membrane structure and electrical properties.

The ability of an applied field to induce membrane molecular reorganization has been demonstrated in several studies. In a pioneering investigation, Poo and Robinson (1977) observed the redistribution of labeled concanavalin A (con A) receptors in muscle-cell membranes in response to an electric field of 4 V cm-1. They noted that con A movement "seemed to be electrophoretic in nature". Similarly, Lee et al. (1994) demonstrated that an applied electric field gradient could induce liquid-liquid phase separation in a monolayer binary mixture of dihydrocholesterol and dimyristoylphosphatidylcholine (DMPC). More recently, Groves et al. (1997) examined field effects on a membrane bilayer. In this more biologically realistic study, a membrane consisting of fluorescently-labeled lipid NBD-PE, cardiolipin, and egg-phosphatidylcholine (egg-PC) was supported on a silica substrate in which a ~10 Å layer of water separated the membrane from the solid surface. The latter feature permitted free diffusion of both leaflets over the surface of the substrate, thereby approximating many of the properties of a natural membrane. Barriers to lateral diffusion ("corrals") were created by scratching the membrane-coated surface. The application of a ~40 V/cm electric field tangent to the membrane plane caused the NBD-PE probe and cardiolipin to "drift toward the anode side of the corral and build up concentration gradients against the barrier". Field strengths in these experiments were much lower than the 100 kV/cm value produced during ion-channel gating. Thus long-range reorganization of membrane microdomains into larger structures consistent with the "coalescence" described by Hao et al. (2001) during permeant ion movement appears likely.

3.3. Membrane Ethenes Are Polarized

The field-induced coalescence of sphingolipid/cholesterol-enriched microdomains into domains or "rafts" may generate an additive alignment of hydrocarbon ethenes. The condensation of sphingolipids in the presence of cholesterol (see discussion in 2.9, above) is consistent with interaction between the ethenes of adjacent hydrocarbon chains (Brown, 1998). Ethene interaction is also enhanced by the cholesterol-induced stability of lipid hydrocarbon chains (i.e., reduction of chain movement) (McConnell and Radhakrishnan, 2003).

Finally, the alignment of lipid ethenes in the plane of the bilayer and tangent to the applied field is energetically favorable. This latter feature was demonstrated in two computational studies conducted by the author and Price (Price and Wallace, 2001; 2003). Because this property is the key feature in the present model of neuron signaling, these two studies will be discussed at length.

3.3.1. First Computational Study: Rationale

In this first study, we wished to see if aligned ethenes could be polarized in an applied field. With that objective, we examined dipole and quadrupole moments and polarizability in adjacent ethenes of a monomer, dimer, and trimer (Price and Wallace, 2001). The electrical permanent moments (dipole, quadrupole) represent derivatives of the energy E with respect to the applied electric field vector $\vec{\xi}$. Specifically, dipole moment (μ) is given as $-(dE/d\vec{\xi})$ and quadrupole moment (θ) is given as ½ $(d^2E/d\vec{\xi})$. Similarly, polarizability (α) is given as $-(d^2E/d\vec{\xi}^{\,2})$ and varies with the oscillation of an applied electric field (Dykstra, 1997). The latter feature is consistent with the complex, interacting fields (with time-dependent variation in strength) that are encountered in actual membranes. (On this point see Kinnunen and Virtanen, 1986). The use of a polymer model stabilized by methylene linkages was justified by computational pragmatics. When ab initio methods are applied to disconnected model components, radical changes in system geometry are the frequent result. We additionally wished to demonstrate that a systematic increase in the number of aligned ethenes would produce a dramatic increase in the observed polarization.

3.3.2. First Computational Study: Method

Our calculations were performed on an ethene monomer, dimer, and trimer using Hyperchem version 4.5 (Hypercube Inc., 1995) and Gaussian 95 W (Frisch et al., 1995). The latter two components were stabilized by methylene linkages. The 4-31G* basis set was used to optimize all structures. Dipole and quadrupole moments and polarizability values were then calculated on the optimized system using the same basis set.

3.3.3. First Computational Study: Results

Sensitivity of the aligned ethenes to an applied electric field was indicated by our preliminary analysis of dipole and quadrupole moments, and polarizability (See Table 1, Appendix). Proceeding from the ethene monomer to the trimer, total dipole moment (μ_{total}) increases from 0.000 to 0.6538 Debye. Examination of the quadrupole moments (θ) reveals a consistent 5-fold increase in the axial moments (i.e., xx, yy, zz) from monomer to trimer. These cumulative increases suggest a progressive mobilization of π electrons. Pronounced fluctuation of the π electronic structure in the applied field is indicated by component increases in polarizability (α) from monomer to trimer (α_{xx} from 7.672 to 93.644, α_{yy} from 19.226 to 125.285, α_{zz} from 30.875 to 91.345 J^{-1} C^2 m^2). These increases in turn produced an increase in mean polarizability $<\alpha>$ from 19.258 to 103.424 [where $<\alpha> = (\alpha_{xx} + \alpha_{yy} + \alpha_{zz} / 3)$]. Together these preliminary results indicate an overall increase in the stability of the aligned ethenes as well as a pronounced increase in the mobilized electron polarizability.

3.3.4. Second Computational Study: Rationale

In our second study, we wished to extend the above results by evaluating the electronic properties (i.e., dipole and quadrupole moments, and polarizability) of aligned and extended ethene model compounds. Based on our preliminary analysis, we anticipated an increased sensitivity to field gradients as the number of ethenes was increased. Second, we expected that the extended-ethene model would reveal the effect of the ethene spatial organization on the relative importance of dipole and quadrupole moments in establishing the extent of ethene interaction with the electric field or the field gradient. Finally, we investigated ethene polarizability in model galactocerebroside molecules.

3.3.5. Second Computational Study: Method

In order to investigate the effects of molecular organization on the electronic properties (i.e., dipole and quadrupole moments, and polarizability) of ethylenic units, extended and stacked model compounds were designed and constructed using Hyper Chem Pro 6 (See Figure 1, Appendix). Initial geometry optimization was performed using the MM + force field. Structure files were converted to Z-matrix form and exported to Gaussian 94W. Geometry optimization was then performed using the 4-31G* basis set. Dipole (μ) and quadrupole (θ) moments and polarizability (α) were calculated using the 4-31G* basis set. Stacked and extended models were designed to demonstrate the effect of ethene alignment believed to be characteristic of microdomains by contrasting the alignment with a more randomized molecular structure. Due to additivity, contributions made by the linker atoms to the quadrupole moment and polarizability were subtracted from the calculated moments to yield a corrected estimate of θ and α Because contributions made by the linker atoms to μ were not significant, values were taken to represent contributions made by ethylenic bonds.

Molecular mechanics and semi-empirical methods were used to examine the effect of ethylenic bond organization on polarizability. Because the investigation sought to determine relative trends, simple clusters containing 16 lipids were constructed using Hyper Chem Pro 6. Geometry optimization was performed using the OPLS force-field with the following constraints: 1) the dielectric constant was scaled to 2.5; 2) a smooth switching function was employed to neglect interactions over distances exceeding 14 Å. Geometries were optimized using conjugate-gradient methods supplied with the program. Optimization was complete when the energy gradient decreased to 0.1 Kcal/mole-Å.

The average polarizability of ethylenic bonds was estimated via a single-point semi-empirical PM3 calculation. The calculation was performed on isolated ethylenic units obtained by selecting all ethylenic bonds, deleting the remaining atoms, and capping each ethene with two hydrogen atoms. The spatial organization of the ethenes was not changed.

3.3.5. Second Computational Study: Results

The properties of molecules in static or oscillatory electric fields influence the strength of intermolecular interactions and the bulk electrical properties of a molecular system. Ab initio methods were used to determine the dipole moment (μ), the quadrupole moment (θ), and polarizability (α) of ethylenic bonds in model systems designed to reflect different levels of organization (See Figure 1, Appendix). Although simple in construction, these models (as noted above) make it possible to investigate how the alignment of ethylenic bonds within microdomains would affect field responsive properties. Stacked and extended models therefore represent organized and randomized states, respectively. Results have been obtained for one, two, and three ethylenic bonds.

The response of a molecule to an electric field can be considered a perturbation, and hence the molecular energy will depend on the electric field strength. The energy (*E*) of a molecule in the presence of an electric field ($\vec{\xi}$, a vector) is related to its total charge (*q*), and the work required to move a charge between two points separated by a distance *r*, the scalar potential (φ). The interaction of the dipole moment ($\vec{\mu}$, a vector) and the quadrupole moment (Θ, a tensor) with an electric field is represented by taking the energy as the sum $E = \Sigma_i q_i\, \varphi(r_i)$ and using a Taylor series expansion for φ, recalling that $\xi = -\nabla\varphi$, the gradient of φ which represents the directional rate of change of φ. Thus, the Taylor series expansion of the energy of a collection of charges interacting with an electric field is

$$E = q\varphi - \vec{\mu}\cdot\vec{\xi} - \frac{1}{3}\Theta:\nabla\vec{\xi} + \ldots.. \qquad (1)$$

Equation 1 illustrates that the energy is a result of the total charge *q* interacting with the potential φ, the dipole interacting with the field ξ, and the quadrupole moment Θ interacting with the field gradient $\nabla\xi$. It is also important to understand that the properties appearing in Eq. 1 represent spatial functions. Thus, as a scalar property, φ has a single value at a given position in space (x,y,z). The vector quantities $\vec{\mu}$ and $\vec{\xi}$, on the other hand, have both a value and a direction at any point in space. The quadrupole moment θ is a quantity that depends on the orientation of the system; hence it is a tensor property.

Just as φ was expanded, it is possible to expand $\vec{\mu}$ in a power series with respect to $\vec{\xi}$. The resulting series is

$$\vec{\mu} = \vec{\mu}_o + \alpha\cdot\vec{\xi} + \frac{1}{2}\beta:\vec{\xi}^{\,2} + \ldots.. \qquad (2)$$

This series reveals that in the presence of an external field, perturbation of the molecular charge leads to new moments. The first term $\vec{\mu}_0$ is the permanent dipole moment, and the subsequent terms represent induced moments. The second term $\alpha \cdot \vec{\xi}$ is linear with respect to the applied field and introduces the contribution of polarizability α. Like θ, α is a tensor property. The third term $\beta : \vec{\xi}^2$ is non-linear with respect to the field and represents a higher-order tensor property known as the hyperpolarizability β. This factor was not considered in this work. As was the case for the series shown in Eq. 1, more terms are possible but were not included in this work. Neglecting β reduces the series in equation 2 to two terms, thus making it possible to equate the perturbed dipole moment to the permanent and induced moments. This yields

$$\vec{\mu} = \vec{\mu}_o + \vec{\mu}_{induced} \qquad (3)$$

where $\vec{\mu}_{induced} = \alpha \cdot \vec{\xi}$. Taken together these equations provide a unified description of how the energy of a molecule interacting with an electric field is coupled to specific interaction terms (Eq. 1), and how these field-dependent interactions indicate the proclivity for an electric moment to be induced by an electric field (Eq. 2). These relationships permit better understanding of how ethylenic bonds in the neural membrane might respond to transient electric fields and field gradients resulting from ion channel activation.

The total dipole moments, μ_{total} for model compounds reveal interesting trends (See Table 2, Appendix). Specifically, μ_{total} increased from 0.000 Debye for a single ethene to 0.536 Debye when two ethenes occurred in a stacked (organized) conformation. Significantly, μ_{total} decreased more than 5-fold to 0.095 Debye when two ethenes assumed an extended (randomized) conformation. This result is consistent with cancellation of dipoles that occurs when two dipoles point in opposite directions. An equally significant result was obtained with the ethene trimer. In this instance, the total moment for the trimer is 1.22 times greater than μ_{total} obtained for the stacked dimer. This result is significant because it reveals additivity, an essential element of many-body systems. In a perfectly additive system, the magnitude of μ_{total} would increase in proportion to the number of dipolar centers. Thus, μ_{total} for the trimer would be expected to be $(3/2)\mu_{total}$ (dimer, stacked) or 1.5 greater than μ_{total} (dimer, stacked) which translates to a μ_{total} (trimer) of 0.804 Debye.

Values for the orientation sensitive property θ are indicated. This higher order term present in Eq. 1 becomes important when the dipole moment is small. The magnitude of μ_{total} for the different model compounds is small compared to that expected for a polar molecule such as water. As a result the contribution of θ would become more pronounced. The results clearly indicate an additive increase in θ upon increasing the number of ethenes from one to three. Specifically, the calculated average quadrupole moment increases in absolute terms from 13.162 Debye-Å, for the ethene monomer, to 24.629 for the stacked dimer, reaching a maximum of 37.299 Debye-Å for the stacked trimer. In sum, this represents a 1.9 to 2.8-fold change relative to the monomer. The 2.4-fold change observed with the extended dimer is consistent with the greater spatial extent of the extended system compared to the more organized stacked system. Taken together, these results indicate an increased sensitivity to field gradients as the number of ethylenic bonds is increased, and an intriguing response to different conformational states (i.e., stacked vs. extended).

Polarizabilities are also indicated (See Table 2, Appendix). Of the electrical properties discussed so far, this one is central to the model because it links the field-induced perturbation of the charge distribution of an ethylenic bond to its ability to interact with the field. Examination of the results indicates that the average polarizability, $\langle\alpha\rangle$, increases in an additive way with respect to the number of ethenes, increasing from 19.258 for ethene to 56.508 $Bohr^3$ for the ethene trimer, a 2.9-fold increase.

In this context it is worth mentioning the calculated properties of ethene dimers in the extended conformation. This model represents a less organized state relative to the stacked conformation. When compared to the latter state, two things become apparent. First, the decrease in μ_{total} upon going from the stacked to the extended conformation is consistent with the compensatory increase in $\theta_{average}$. This finding supports our supposition that spatial organization of ethylenic bonds within a microdomain dictates which field-dependent term in Eq. 1, μ or θ, will be important in establishing to what extent an ensemble of double bonds will interact with the electric field $\vec{\xi}$ or the field gradient $\nabla\vec{\xi}$. Second, and perhaps most significant, the results suggest that interaction and responsiveness to transient electric fields increase as the number of ethylenic bonds increases. This has profound implications since properties of many systems are extensive, i.e., they depend on the number of components. Two familiar examples are the collective strength of hydrogen bonds within a protein, and double-stranded DNA.

The next stage of this investigation used molecular mechanics and semi-empirical methods to determine if ethene polarizability (α) in a cluster of lipids exhibited trends paralleling those observed in the simple ethene-containing model compounds. Three model systems were constructed (See Figure 2, Appendix).

The Type 1 cluster contained 16 tightly-packed nervonic-acid-containing galactocerebroside (NervC) molecules. The Type 2 cluster contained 16 ordered, but less densely packed, NervC lipids. In the third model system, the Type 3 cluster contained a randomized mixture of 8 NervC lipids and 8 steric-acid-containing galactocerebroside lipids. All clusters were energetically favorable, having energies ranging from –630 Kcal/mol for Type 1 cluster to –441 Kcal/mol for the Type 3 cluster. The energy of the Type 2 cluster fell approximately midway between the above two values (–503 Kcal/mol). Molecular models of these clusters are indicated. It is interesting that the head group ethenes and those located in the nervonic acid chain form wire-like assemblies in the Type 1 and 2 clusters. In contrast, these ethenes appear loosely organized and randomized in the Type 3 cluster. The average polarizability $<\alpha>$ appears relatively constant when the head-group ethenes are compared to those residing in the nervonic acid acyl chain. Significantly, the results reveal additivity. Specifically, the polarizability of acyl chain ethenes in the mixed Type 3 cluster is approximately one-half as large that calculated for those in Type 1 and Type 2 clusters. This is an expected result, since the Type 3 cluster contains 8 NervC molecules and the other clusters contains twice as many.

In summary, the field applied in the plane of the bilayer during ion movement through an open channel reorganizes the membrane such that sphingolipid microdomains coalesce into domains. The concentrated sphingolipids contain unsaturated bonds in an energetically favored arrangement aligned parallel to the plane of the membrane and in the direction of the applied field. The aligned ethenes become highly polarized such that the electropositive end of each dipole is closest to the source of the field (i.e., ion movement through the open ion channel).

3.4. Membrane Dipoles Interact Electrostatically with Ion-Channel Charge Centers

Aligned and polarized ethenes in the membrane are hypothesized to electrostatically interact with charged amino-acid residues in an unfolded ion-channel protein (See Figure 3, Appendix). The duration of the interaction, $T_0 + \Delta T$, where channel gating is T_0, is proposed to establish the open time of the channel T_{open}. (The computational properties of this system will be discussed in

Chapter 5). *The interaction is the proposed electrochemical basis for neural membrane regulation of neuron signaling.* Recent *in vitro* studies of lipid-bilayer folding systems for membrane proteins suggest a close proximity of membrane lipids to protein α-helices consistent with the proposed electrostatic interactions (Booth and Curran, 1999; Booth et al., 2001; Liang, 2002; Cantor, 1999). The close proximity is primarily the result of membrane lateral pressure due to reduction of an energetically-favored monolayer curvature. Booth and Curran (1999) have noted that "most biologically relevant lipids form structures in which the polar/apolar interface of the monolayer wishes to bend toward the water". When the monolayers are back-to-back in a bilayer, a torque tension arises due to the stored curvature energy of each flat monolayer. "A protein folding [e.g., ion-channel closing] in the bilayer will feel the effect of these various lipid forces". Lipid-protein interaction may additionally be enhanced by the conformation of the channel protein itself. Liang (2002) has noted that membrane proteins "contain numerous voids and pockets" (packing defects) which provide spaces for "binding ligands, prosthetic groups, lipids and water molecules, and facilitate conformational changes essential for protein function". Finally, the assumption that membrane lipids and the ion-channel protein are in close juxtaposition is consistent with a tryptophan-scanning study of the voltage-dependent *Shaker* K^+ channel conducted by Hong and Miller (2000). In this study, it was found that the S1 segment of the channel has a "well-defined lipid-exposed surface running along its entire length across the membrane." Additionally, the helical region of the S3 segment (~20 Å in length) is exposed to membrane lipids on half of its surface.

Additional biochemical evidence suggests that unfolded membrane proteins, including voltage-gated ion channels, contain positively and negatively charged transmembrane amino acid residues critical to voltage sensing, ion conduction, channel closing, sensing of H^+ concentration, and binding of extracellular substances (Tiwari-Woodruff et al., 1997; Torshin and Harrison, 2001; Madcja, 2000; Catterall, 2002). Of particular relevance is a model of electrostatic interactions in 20 small proteins developed by Torshin and Harrison (2001). The model, which was based on protein-sequence and 3D-structure databases, proposed that attraction between negative residues (aspartates and glutamates) and positive residues (arginines and lysines) in the hydrophobic core of the protein triggered an annealing-like "collapse" (minimum local-potential-energy search) to the compact, native state. Similarly, Madeja (2000) has summarized recent x-ray crystallography, hydrophobicity-profile, and site-directed mutagenesis studies indicating that voltage-gated channels contain surface charges in the extracullular medium essential to voltage sensing, substance binding and sensing of H^+ ion concentration. These charges include the negative (aspartate and glutamate) and positive (arginine and lysine) residues noted by Torshin and Harrison (2001) as well as positively-charged histidine residues. Considered together with the computational model discussed above for transient neural-membrane dipoles, these findings suggest that electrostatic interaction between ion channels and the surrounding membrane is a plausible physical basis for ion channel regulation and, hence, neuron signaling.

3.5. The Reset Mechanism: The Reversal of Field Effects

The membrane-regulated conformational change in the ion-channel protein from the random coil (open-pore) to the α-helix (closed-pore) structure causes cessation of ion flow and the removal of field effects from the membrane. In the absence of the field, the membrane lipids and cholesterol return to a state of minimum local potential energy. On thermodynamic grounds, the probable molecular arrangement would be that of the initial state prior to electrochemical perturbation (Li et al., 2001; Mitchell and Litman, 1998; Wang et al., 2000, 2001; Samsonov et al., 2001; Brown, 1998; Hao et al., 2001). Thus, cholesterol would again occupy its former energetically-favored location adjacent to sphingolipids. In this location, the compact, planar tetracyclic rings of cholesterol would once more be adjacent to the hydrocarbon chains of the sphingolipids. This favored interaction would re-create microdomains separated by an interfacial region of polyunsaturated lipids.

The two most recent studies of membrane field effect are consistent with the "reset" feature of the model. Groves et al. (1997) noted in their study of electric field effects on a two-component planar supported lipid bilayer that removal of the field causes the membrane to "relax back to uniformity" by diffusive mixing of components. Similarly, Radhakrishnan and McConnell (2000) found that removal of the applied field causes the re-formation of condensed complexes of phospholipids and cholesterol in a monolayer. Clearly, more studies of membrane relaxation following field removal are needed. At this point, it intuitively seems likely that the time required for a destabilized membrane to return to the unperturbed, liquid-ordered state (Hao et al., 2001) would significantly affect the duration of the ion channel "open" times during successive depolarizations.

Chapter IV

Membrane Microdomains and Neural Impulse Propagation: Field Effects in Cytoskeleton Corrals

The possibility that an ensemble of neural-membrane lipids could regulate the duration of the ion-channel "open" conformation may have direct implications for local gating of the action potential (AP). The control of channel dynamics is equivalent to controlling the transmembrane ion conductances responsible for neural depolarization and hyperpolarization. Accordingly, if a membrane region in an axon were to contain (for example) clusters of Na^+ channels with relatively longer open states, such a structure would be conducive to a local depolarization (spike) and the continuing propagation of the neural impulse. By contrast, if the Na^+ channels had brief open states, the structure would be conducive to impulse propagation failure. Modulation of the membrane state through other channel types would of course also be possible. A good example is the A-current delayed-rectifier K^+ channel (K_A), currently the object of intensive research because of its possible role in conduction failure. The K_A channel is gated by membrane hyperpolarization, which it increases and prolongs by rapidly conducting potassium ions out of the cytosol. A prolonged open time for this channel could hypothetically strengthen the delayed-rectifier effect and significantly increase the probability of conduction block.

This chapter examines the possible role of neural membrane microdomains in regulating the propagation of the action potential. These data are particularly important because they appear consistent with the concept that "local switches" regulate AP propagation (Scott, 1995). The chapter begins with an overview of the experimental evidence for AP conduction failure. It then examines the two major alternative models: one emphasizing the role of impedance mismatch due to neuron branching geometry; the other emphasizing the role of prolonged hyperpolarization due to the A-current potassium channel (K_A). A model emphasizing the possible interaction of a microdomain-cytoskeleton system with neuron branching geometry will then be presented (Wallace, 2004).. Because of the large number of studies bearing on the subject, the K_A channel will be used as the basic example, although the possible contributions of other channel types will be briefly discussed. The chapter concludes with a discussion of how microdomain regulation of AP propagation may explain a number of neuron features that strikingly depart from cable properties.

4.1. Conduction Failure in Neurons: Evidence and Models

The earliest evidence for AP propagation failure in neurons preceded by 17 years the patch-clamp experiments conducted on the squid axon by A.L. Hodgkin and A.F. Huxley that culminated in the standard model of neuron electrical signaling (1952a, 1952b, 1952c, 1952d). Thus it was evident nearly 70 years ago that a neural depolarization (spike) was not inevitably propagated from the point of its origin to the presynaptic terminal. D.H. Barron and B.H.C. Matthews (1935) applied electrical stimuli to a cat spinal cord axon and recorded the propagating signals at two locations. They found that signals recorded at the first point were intermittently not detectable at the second point, but the physical basis for the conduction failure was unclear.

Following the Barron-Matthews study, axonal propagation failure was experimentally demonstrated in a variety of vertebrate and invertebrate species: e.g., the molluscan central neuron, crayfish abdominal axons, the walking leg of the crayfish, leech sensory neurons, rat motoneurons, and the visual callosal axons of the rabbit (Tauc and Hughes, 1963; Parnas, 1972; Hatt and Smith, 1975; van Essen, 1973; Krnjevic and Miledi, 1959; Swadlow and Waxman, 1976). Similarly, dendritic conduction failure was identified in cochlear neurons of the monkey, hedgehog, owl, and bat, and in alligator Purkinje neurons (Bogoslovskaya et al., 1973; Llinas et al., 1969). As the data accumulated, it became evident that axonal and dendritic branch points frequently had a low "safety factor" for impulse conduction. However, there was (and is) no consensus explanation. Instead there has been the largely separate development of models emphasizing either branching geometry or local hyperpolarization driven by the A-current potassium channel. The possibility that a molecular ensemble of some type could "decide" whether or not AP propagation should continue has not been systematically explored.

In the former set of models, impulse failure is related to the degree of impedance mismatch associated with the diameters of mother and daughter neural branches. The optimal geometry is given by Goldstein and Rall (1974) as

$$d_0^{3/2} = d_1^{3/2} + d_2^{3/2} \tag{1}$$

where d_1 and d_2 are diameters of daughter branches and d_0 is the parent branch radius. Accordingly the Goldstein-Rall (GR) ratio is given by

$$\mathrm{GR} = d_1^{3/2} + d_2^{3/2} \,/\, d_0^{3/2}. \tag{2}$$

When GR is 1, impedances are perfectly matched and the conduction probability is high.

For GR< 1, the AP propagates with slightly increased velocity, as if the axon were tapering (Koch, 1999). The most common situation is GR > 1, where the electrical load of the daughter branches exceeds that of the parent branch. For 1<GR<10, conduction past the branch point is still assured, with the failure rate increasing for GR>10. Applications of the GR ratio to actual neurons have encountered mixed success. Rall successfully applied it to the electrical behavior of cat α motoneurons in 1977 (see also Rall et al., 1992) inspiring a related study by Fleshman, Segev, and Burke (1988). However, as several researchers subsequently determined, real neurons seldom conform to the structural requirements of the GR model: e.g., the 3/2 diameter rule at branch points, and "the assumption that all dendrites terminate at the same distance from the cell body" (Koch, 1999).

Alternative models of AP conduction failure have emphasized the role of the A-current potassium channel (K_A). The K_A current is an outward transmembrane flow of K^+ ions activated by hyperpolarization following a depolarizing pulse (Levitan and Kaczmarek, 2002). The current is subsequently inactivated at the return of the resting potential (-60mV). The K_A channel, by prolonging hyperpolarization, slows the return of the transmembrane voltage difference toward the resting potential. Clusters of K_A channels at branch points theoretically maintain the interior of the neuron at a hyperpolarized potential for a sufficient period of time to block the propagation of an AP This possibility has been investigated through *in vitro* slice electrophysiological studies and computer simulations (Debanne et al., 1999; Kopysova and Debanne, 1998). It was found that a K_A current could be activated by a brief (150 msec) hyperpolarizing prepulse (where voltage command ranged between -50 and -110mV) followed by a somatic depolarization (+30mV) of 50 msec duration.

The geometric and K_A approaches effectively frame the issue: What are the relative contributions of neuron branching geometry and branch-point K_A channel clusters to AP conduction failure? The query raises a related---and more general---question: Do K_A and other types of channel clusters communicate with one another, thus constitituting a molecular network operating within geometric constraints?

4.2. Regulated Lipid Diffusion through Cytoskeleton Gates

Viewed from an adaptive standpoint, a neuron must be able to modify its AP propagation behavior with regard to fluctuating cognitive or behavioral requirements. This is perhaps the most plausible argument for a synthesis of the ionic and geometric approaches: micro-structural changes in the ion-channel environment can alternately reinforce or overcome the effects of neuron geometry. For example, neocortical neurons are components of neural networks subserving integrative cognitive functions. Accordingly, these neurons are characterized by highly arborized axons (Destexhe et al., 2003). However, in terms of the GR model and related geometric approaches, the branching geometry, while necessary for integrating cognitive modules, is highly unfavorable for AP propagation success. Accordingly, a molecular-level system must overcome this structural problem by insuring high depolarizing spike amplitude (and propagation success) at the axonal branch points. Conversely, in hippocampal CA3 networks, which according to several models, simplify or "index" complex cortical inputs for cross-modal comparison and identification of novelty, the suppression of irrelevant information would be essential for adaptive function (Gray, 1982. Also see review of comparator models in Vinogradova, 2001). Axonal branch points of CA3 neurons should thus be enriched with molecular systems that selectively reinforce the effects of neuron branching geometry by generating prolonged membrane hyperpolarization.

The modulating effects suggested above would require stepwise adjustments in the concentration of membrane unsaturated lipids proposed in this study to regulate ion-channel dynamics. Of course, if the membrane were indeed a fluid mosaic, as originally suggested by Singer and Nicolson (1972), the stochastic mixing of lipids would make such a regulatory role impossible. For that reason, the possible existence of gated barriers regulating local membrane lipid composition was speculatively suggested by Kinnunen in 1991. At that time, however, there was little or no evidence that such barriers actually existed. This situation is rapidly changing as new evidence is being put forward indicating that the neuron cytoskeleton may be equipped with dynamic molecular gates.

The cytoskeleton is a dense protein matrix of microtubules, intermediate filaments, and actin filaments located immediately beneath (and to some extent, within) the membrane in all eukaryotic cells (Kirkpatrick and Brady, 1999). The structure is the basis for the unique morphology of cells comprising different types of tissue; without it, an animal cell would be an amorphous bag of fluid. In addition, the cytoskeleton defines metabolic compartments within the cytosol, and provides tracks for intracellular transport. Recent evidence suggests that cytoskeleton compartments ranging in diameter from 100-600 nm and with a thickness of ~6 nm enclose membrane lipids and regulate their inter-compartmental diffusion in a stepwise fashion (See Figure 4, Appendix). The earliest investigations suggesting regulated lateral membrane diffusion focused on proteins rather than lipids. The movement of protein band 3 within the erythrocyte membrane was the basis for computational models that utilized a Monte Carlo approach (Tomishige et al., 1998; Tomishige et al., 1999; Brown et al., 2000; Leitner et al., 2000). The Tomishige studies found that, on average, a Band 3 protein moves from one cytoskeleton compartment to another every 350 ms. The investigators proposed that intercompartmental diffusion may be regulated by spectrin, a component of the cytoskeleton. The spectrin molecule consists of two chains which are aligned in an anti-parallel arrangement and wound around each other to form a heterodimer. Two heterodimers then associate head-to-head to produce a tetramer ~200 Å in length. The spectrin network is attached to the membrane through binding with ankyrin and protein 4.1 (Vale et al. in Hall, 1992). In the "skeleton fence" or "picket fence" cytoskeleton model, the spectrin heterodimer dissociates into two dimers, and then re-associates into a tetramer, thereby creating a dynamic gate for intercompartmental molecular diffusion.

The dissociation of the protein tetramer can be "fast" or "slow". Average opening rates, as determined by the Monte Carlo studies, were 5, 10, and 20 s^{-1}, while closing rates ranged from 1 to 10^6 s^{-1}. Accoiding to the Tomishige model, the increasingly faster closing rates would confine lipids inside a compartment. However, the precise physical basis for spectrin dissociation-reassociation remains controversial. Separate studies identify chemical and electrical mechanisms, and it may be that both are required (Wallis et al., 1992; Wallis et al., 1993; Zagon et al., 1986; Vassilev et al., 1982; Vassilev et al., 1983; Vater et al., 1998). Brain spectrin conformational change in response to Ca^{2+} binding to four high-affinity sites has been demonstrated by flow dialysis and NMR studies. However, the process by which intracellular Ca^{2+} concentration is modulated during neuron signaling, i.e. capacitative calcium entry through the endoplasmic reticulum, is not well understood (Putney, 2003). Alternatively, microtubules, which directly cross-link with spectrin in vivo, have been observed in solution to become aligned in an electric field because of the microtubule internal dipole. It follows, at least hypothetically, that field-induced microtubule alignment during an AP could mechanically modify the conformation of bound spectrin.

Do lipids move through cytoskeleton gates in a manner comparable to protein Band 3? The Kusumi Membrane Organizer Project is presently investigating this process through single-particle tracking studies (Fujiwara et al., 2002; See review in Kusumi, 2005). In their 2002 study, the movement of the unsaturated (and *non-microdomain*) lipid, 1,2-dioleoyl-sn-glycero-3-phosphoethanolamine (DOPE), labeled with Cy3, a long-wavelength dye frequently used in fluorescence microscopy, was tracked in a normal rat kidney fibroblastic cell membrane. The study found that the labeled lipids moved between cytoskeleton compartments via hop diffusion. Compartment size was ~230 nm, and the mean residency time for the lipids was 11 ms. Although these findings appear to be consistent with the present model, they must be viewed with caution. DOPE, as noted, is a non-microdomain lipid. Thus, the hop diffusion rate for microdomains is unknown. Also, the NRK membrane contains double (or nested) compartments; i.e. it is compartmentalized into 750-nm and then into 230-nm compartments. This type of compartmentalization has not been found in other cell types, including neurons. However, shortly after this study, lipid diffusion was investigated in the neuron utilizing both fluorescently-labeled phospholipids and lipids attached to gold particles (Nakada et al., 2003). As Boiko and Winckler (2003) note in their commentary, each technique has its limitations. Weak and short-lived fluorescence signals generate a low signal-to-noise ratio, and consequent low spatial and temporal resolution. On the other hand, the results of single particle tracking studies may reflect the interaction of the attached bead with extra-cellular components. Nonetheless, in the Nakada study, similar results were obtained through both techniques. Diffusion of membrane lipids out of the axon initial segment (AIS) was blocked; lipids were immobilized in the AIS region. Importantly, ankyrin binding proteins (proposed components of cytoskeleton fences) are enriched in the AIS, consistent with the picket fence model.

4.3. The K_A Potassium Channel and Neural Impulse Regulation

The evidence presented above appears consistent with a molecular regulatory system comprised of corralled microdomains and the K_A channel operating in concert with branching neuron geometry to control the propagation of the action potential (See Figure 5, Appendix). The basic properties of the proposed molecular system are the following: 1) K_A channels are distributed in clusters at neuron branch points and axonal swellings, for which there is a low safety factor for AP propagation. 2) An individual K_A channel or subset of channels within a cluster can be activated by an electric field conducted in the plane of the membrane. 3) Individual K_A channels are surrounded and stabilized by microdomains. 4) Microdomains vary in saturated/unsaturated lipid ratios. 5) K_A channel clusters and lipid microdomains are enclosed by a cytoskeleton corral. 6) Cytoskeleton corrals are dynamically gated by spectrin tetramer dissociation and reassociation. 7) Average gating rates vary between corrals, with faster gating rates confining lipids within the enclosure. Together these features may comprise a molecular regulatory system capable of rapid on-line modulation of the AP neural code.

As an example of how this system may operate, we may first consider a K_A channel cluster associated with a high rate of conduction failure, as in hippocampal CA3 neurons discussed above. In this type of molecular system, K_A channel clusters would be associated with unsaturated-lipid microdomains and slow cytoskeleton gates. The applied electric field associated with K_A channel gating would induce intra- and intercompartmental diffusion of unsaturated lipids toward the source of the field. As a result, the concentration of unsaturated lipids would increase in the immediate vicinity of the channel. The adjacent unsaturated bonds, both proximate to the head group and at the kink in the diacyl chains, would become aligned and polarized in the applied electric field. The folding of the gated K_A channels to the closed α-helix conformation would be "timed" by electrostatic interactions between charge groups in the unfolded (random-coil) channel and dipoles in the surrounding membrane. The relatively large number of electrostatic interactions involved would constitute a form of complex problem, requiring a protracted search for a solution (the lower-energy closed-channel conformation). In this manner, an elevated concentration of unsaturated lipids would prolong the open time of the gated K_A channels. Prolonged K_A channel open time would in turn generate strong local hyperpolarization, increasing the probability of AP propagation failure.

Alternatively, we may consider the function of corralled K_A channels in branching neocortical neurons, where AP propagation through branch points (i.e. low failure rate) is essential for normal network activity. The present model predicts that branch points in neocortical neurons would be enriched with K_A channel clusters associated with saturated-lipid microdomains and fast-gated cytoskeleton corrals. In this type of system, relatively brief K_A channel open times would result from the reduced complexity of the electrostatic interactions. This feature would reduce local hyperpolarization and thus increase the probability of AP propagation success. It might be argued that AP propagation success at neocortical neuron branch points could be more directly achieved by the absence of K_A channels altogether. Although this argument has an intuitive appeal, the complete absence of K_A channels would likely not be adaptive. In many systems of neurons, and neocortical neurons in particular, it would be physiologically essential to have an "emergency" system to dampen uncontrolled activity (e.g. epilepsy). This aspect of K_A channel behavior will be examined in some detail in the following chapter.

4.4. Implications of the Model: Molecular Networks in the Neuron

It has been recently suggested that the existence of cytoskeleton corrals enclosing membrane lipids may constitute a "paradigm shift" in cell biology (Kusumi et al., 2005). The present study would extend that claim by proposing that the new viewpoint may introduce a similar shift in our understanding of nervous systems as well. The K_A channel is the best-studied example of several types of ion-channel clusters which could communicate with one another and thereby regulate neuron signaling in complex and subtle ways. It is a commonplace that neurons are organized into networks. But are molecular networks operating within each neuron? Whatley and Harris (1996) have reviewed a large number of studies indicating that several types of ion channels interact directly with the cytoskeleton and are distributed in clusters. Examples include the ligand-gated channels such as nicotinic acetylcholine receptors, glycine receptors, glutamate receptors and $GABA_A$ receptors. In addition, Rasband and Schrager (2000) observed Na^+ channel clustering at nodes of Ranvier, and Shaker-type K^+ channel clustering near axoglial junctions. If subsequent studies reveal that these clusters are part of a larger compartmentalized structure similar to K_A channel cluster organization, the "network" concept would apply to coding within the neuron.

It is intriguing to consider the operation of hypothetical sub-neural molecular networks in relation to neural activity that appears to depart markedly from cable properties. Possibly the most striking example of this departure is an experiment conducted by Magee and Cook (2000) in which excitatory postsynaptic potentials (EPSPs) recorded in dendrites of the hippocampal CA1 region grew "progressively stronger with distance from the cell body, almost exactly counteracting the distance-dependent signal attrition one would expect to find" (Mel, 2002). As Mel notes, this finding is at variance with electrical conduction in a standard cable, where signals typically decay with distance from the source ("voltage drop"). Magee and Cook state that although the mechanism underlying increased EPSP amplitude is presently uncertain, an increased number of AMPA receptors or quanta of released glutamate (or both factors) may be responsible. Either mechanism would appear consistent with the present model. In a related investigation, Hering et al. (2003) demonstrated that AMPA receptors are associated with cholesterol/sphingolipid microdomains, which are necessary for their stability and hence for the maintenance of synapses. Sphingolipid clustering, followed by unsaturated-bond alignment and polarization could regulate the duration of the AMPA receptor channel "open" state. Synchronized "open" states of several AMPA channels would generate a pronounced dendritic spike amplitude at a distal site from the soma. Microdomains may also regulate distal spike amplitude through their possible role in transmitter exocytosis. Salaun et al. (2004) have recently reviewed a growing body of studies suggesting that the SNARE protein complex, essential for vesicle fusion with the neuron plasma membrane, is associated with sphingolipid-rich microdomains. The latter spatially concentrate the SNARE proteins at defined sites, a mechanism which could plausibly function to increase the quanta of released transmitter. Together these examples provide tantalizing glimpses of communicating membrane modules which may regulate AP propagation as well as the strength of the signal itself.

Chapter V

Toward Membrane Molecular Machines: Implications for the Study of Neural Disease

One of the most visible and exciting trends in contemporary biophysics is the artificial mimicry of the highly organized molecular systems found abundantly in nature. This new research direction is ultimately anchored in the optimizing tendency of organic evolution. As biologist Richard Dawkins (1986, 1997) has noted in several popular writings, and as engineer Wirt Atmar (1994) has more formally described, the sustained operation of natural selection on the structures of living systems---the work of the "blind watchmaker"---can result in phenotypes of very high adaptive efficiency. The structures are encountered at several orders of magnitude, including the molecular level. Through a process of "reverse engineering", analysts are attempting to determine the components and movements of highly organized molecular systems, so that their essential features can be artificially synthesized. The ultimate objective is the design of molecular machines. K. Eric Drexler (1994), a pioneer in this effort, notes that "besides producing energy, molecular machine systems can also process matter and information. Among these processing functions, mechanosynthesis is of basic importance because it can be used to build molecular machine systems able to perform the other functions."

This chapter examines the possibility of developing a molecular machine which could possibly simulate the fundamental features of neural-membrane microdomain regulatory behavior. A first-approximation supported-membrane design is outlined. The proposed architecture would consist of an addressable biotemplated nanowire orthogonal to a supported membrane comprised of mixed saturated and unsaturated lipids (Wallace, 2007). Ethene polarization in the applied field of the nanowire would produce electrostatic interaction between the deformable nanowire and the polarized ethenes. This feature would mimic electrostatic interaction between the charge residues of the open ion channel and the polarized ethenes in the surrounding membrane in a living neuron.

In addition, this chapter explores the possible medical usefulness of the proposed design. Although the use of neural-network or artificial-intelligence (AI) models to understand brain disorders is by no means new and is rapidly growing---see for example the recent use of artificial neural networks to distinguish neuropathological lesions of Alzheimer's patients from controls (Grossi et al., 2007)---it is also the case that most of these approaches exemplify classical neuroscience; i.e. they derive from the McCullough-Pitts cable model of the neuron (1943). In this approach, the neuron is construed as a single switch comprised of a unit which calculates a nonlinear sum of inputs with reference to a threshold, as a basis for a firing decision. Although this approach has been, and continues to be, a very powerful and medically useful brain modeling strategy, it may be useful to consider a complementary molecular approach. This possibility is examined at length with reference to the A-current potassium channel (K_A) described in the previous chapter, a channel which is microdomain-localized, and believed to play a role in the etiology of epilepsy.

5.1. Artificial Model: Metallized Nanowire in a Supported Membrane

Based on the evidence presented in previous chapters, it is suggested that the following features be present in the proposed model of membrane-microdomain/ion-channel interaction: 1) a compartmentalized membrane lipid bilayer 2) metallized conducting molecular nanowires, one in each compartment, situated orthogonal to the membrane and sufficiently flexible to be transiently deformed by electrostatic interaction with polarized membrane ethenes 3) a photoresist corral separating each nanowire from the surrounding lipids (thus preventing direct contact of the lipids with the nanowire) 4) a larger corral of vertical carbon nanofibers surrounding the lipid membrane compartments. (In the system schematically depicted [See Figure 6, Appendix] only two compartments are shown). 5) Optically activated gates (molecular rotors), one in each corral, attached to vertical carbon nanofibers.

In preliminary experiments, the objective would be to observe the effects of field-induced membrane lipid reorganization and ethene polarization on electrical conductivity through the metallized nanowires in varying membrane preparations. The following experimental outcomes are predicted: 1) The applied electric field of the conducting nanowire will induce diffusion of unsaturated lipids toward the nanowire. 2) The ethenes of the unsaturated lipids will align in the plane of the membrane and become polarized in the applied field 3) Electrostatic interaction between the conducting nanowire and the polarized ethenes will induce nanowire conformational change (deformation) which will transiently break the flow of current (See Figure 7, Appendix).. 4) The interruption of current will induce a reset mechanism in which membrane ethenes will become nonpolar and the nanowire will relax to its original (conducting) conformation. A nanowire in an unsaturated-lipid environment will thus display a series of pulses in which the inter-pulse intervals are timed by nanowire deformation, thus modeling the channel open state. Conversely, the channel closed state is modeled by continuous transmembrane current, predicted for a nanowire in a saturated-lipid environment.

The selection of possible model components involves the traditional compromise between realism and feasibility. In the case of membrane mimics, there has been progress in developing ~50 Å thick membranes situated on solid supports (Groves et al., 1997; Groves et al., 1998). The support structure typically consists of a silicon dioxide wafer (SiO_2) which interacts with the membrane through electrostatic and van der Waals forces. Lateral fluidity of the membrane leaflets is made possible by a thin (~10 Å) film of water which separates the membrane from the substrate. The corral (cytoskeleton mimic) in the proposed device is comprised of vertical carbon nanofibers (Doktycz et al., 2003). These structures are characterized by multiple layers of catalytically-grown carbon with diameters ranging from ~80-200 nm to that of conventional carbon fibers (~5-10 mm). Recent controlled-growth studies utilizing a Si substrate with Ni-Fe as a catalyst produced a "forest" of vertical nanofibers randomly spaced within the catalyst stripe. Importantly, nanofiber forests can be embedded in a fluidic environment. A small photoresist barrier would surround the nanowire where it crosses the SiO_2 substrate. While permitting field effects, the barrier would prevent physical interaction between the nanowire and the membrane lipids. The proposed corral-gate mimic would exploit light-induced changes in molecular conformation. One possibility is a light-driven molecular rotor in which two different wavelengths would induce reversible rotation of one part of the molecule around a central axis (Feringa et al., 2002). A dynamic molecular gate would thus be a modified version of optical switching devices already in existence in which the rotor is linked to the stator by a C=C axis. In this design, it would be necessary to attach the stator to a single functionalized single-walled carbon nanotube (SWNT), the functional equivalent of a gatepost, located at the nanofiber corral opening. Finally, the choice of an ion-channel mimic involves several possibilities, each having limitations and advantages (Cragg, 2002; Gokel et al., 2004; Biron et al., 2004; Buldum and Lu, 2003; Dai, 2002; Rochefort et al., 1999; Maiti et al., 1996; Dresselhaus et al., 1996). In general, a biomolecule bound to an inorganic compound is preferable to SWNTs or more complex artificial channels such as cyclic peptides, stacked crown ethers, or gramicidin. The SWNT is indeed electrically addressable with a current strength that could in principle be adjusted to within neuron range (1-20 pA) either by doping or adjusting chirality. Nonetheless, its rigidity (Young's modulus $\approx 10^3$ Gpa) would make it unsuitable for functional deformation in a biological system. Additionally, there are obstacles involving STM measurement of SWNT properties, and the present lack of a complete explanatory model for any method of catalytic synthesis. Synthetic ion channels present a different set of problems. Although artificial ionophores such as stacked crown ethers and cyclic peptides have been embedded in membranes and observed to selectively conduct ions, the best transport rates achieved so far are $\sim 10^4$ ions/s (to be compared with 10^6-10^8 ions/s rate of an actual ion channel). Moreover, because the artificial channels are addressed by ion release in an aqueous environment, they are not as precisely controllable as molecular nanowires addressed through electrical inputs. Finally, no method has yet been developed to synthesize an artificial ion channel which can reversibly alter its molecular conformation and ion-transport rate in response to the lipid membrane electrostatic environment.

In contrast to these limitations, nanowires constructed from biomolecules combined with an inorganic compound that acts as a conductor appear promising for nanoscale modeling, although important problems remain to be solved. Recently, a biotemplated nanowire was constructed from a self-assembled amyloid protein fiber in which cysteines were used as binding sites for gold particle (Scheibel et al., 2003). Results were mixed but encouraging. Current density and field strength values were significantly lower than those recorded in the field-induced membrane concentration gradient experiments, and far lower than those recorded in an actual neuron. The maximum values were, respectively, 3.82×10^3 A/cm^2 and 0.125 V/cm for a wire diameter of ~100 nm, an applied voltage of 25 X 10^{-6} V, and an electrode gap width of ~2.0 X 10^{-6} m. The field strength is to be compared with the 4 V/cm value recorded in the con A protein receptor experiment, and the 100 kV/cm value for the neural membrane at the peak of the action potential. Future improvements in conductivity, field strength, and fiber elasticity seem plausible based on the versatility of the system. For example, wire diameter (80-200 nm) can be reduced (thus increasing fiber elasticity) by eliminating silver and gold enhancement between the gold particles. In addition, it should be possible to increase the number of metal binding sites by engineering additional surface-accessible residues such as cysteines and lysines. In the ideal amyloid fiber utilized in a supported-membrane system, electrostatically-induced mechanical deformation would produce a gap between the conducting particles, thus leaving an elongated bare fiber, for which resistance is very high ($R>10^{14}\Omega$).

5.2. A Role for Artificial Membranes in Modeling Neural Disease

Based on the above discussion, it appears possible that molecular computers which model critical features of microdomain regulation of ion-channel activity will emerge in the foreseeable future, and that such models may be useful in understanding the molecular basis of at least some neural diseases. The first possibility clearly depends on one's definition of a computer. Following Churchland and Sejnowski, a computer in the most general sense is a system in which the physical states represent the states of some other system. More precisely, there is "an appropriate (revealing) mapping between the system's physical states and the elements of the function computed" (Churchland and Sejnowski, 1992). In the case of the membrane microdomain, the time values for electrostatic interactions between charge residues of a gated ion channel and polarized membrane ethenes are mapped to changing values of the membrane electrical potential. In addition, it appears possible that the physical states of the system are an input-output transform in which output is a function of input, and (in the case of novel input) the output values converge toward some target (optimum) value. The physics of this learning feature consists of changes in the lipid diffusion gradient between adjacent cytoskeleton corrals in response to changes in the activity pattern of a set of ion channels. Rapid on-line inter-compartmental adjustments of this type would appear to be consistent with the scenario proposed by Koch and Segev in which a single neuron's hardware modifies its own inward and outward currents in response to changes in the sensory environment (Koch and Segev, 2000).

The computations will, of course, go awry if the hardware components are flawed. From the standpoint of disease etiology, component flaws are to be identified with the molecular bases of neural pathologies. Artificial molecular models might therefore be useful for preliminary development of hypotheses to be ultimately tested against living systems. A good example for illustrating possible membrane models is the family of K_A channels discussed earlier, which are believed to play an important role in epilepsy (Birnbaum et al., 2004). The K_A channels are transiently activated when a neuron is depolarized following a period of hyperpolarization. The outward flow of potassium ions through an activated K_A channel slows the return of the membrane electrical potential toward the spike threshold. K_A channels therefore function to decrease AP frequency and reduce the spread of excitability. This function appears to be frequency-dependent, with increasing AP conduction failure occurring at bursting activity above 100 Hz. K_A channels thus constitute low-pass filters. As noted earlier, K_A channels are clustered and microdomain-localized (Martens et al., 2004; Antonucci et al., 2001). Moreover, their dynamics are modulated by alterations in microdomain lipid composition (Martens et al., 2000). The possible medical significance of the latter finding was suggested by a recent study indicating a "potent neuroprotective effect of PUFAs associated with PUFA-induced blockade of glutamatergic transmission." (Lauritzen et al., 2000). Although the spatial pattern of activity within a K_A channel cluster has not yet been investigated, there is a single study of Na^+ cluster dynamics during AP propagation (Savtchenko et al., 2001a, 2001b). Images of active ion channel clusters stained with voltage-sensitive dyes were synchronized with whole-cell recordings. It was found that in each Na^+ channel cluster only a fraction of channels were active during the AP. Moreover, the active fraction varied from trial to trial. Together these data suggest a potentially testable and (perhaps) medically significant model of excitability control.

The proposed experiments would resemble earlier studies involving living cells in which integral protein conformation and ion-channel activity were regulated by experimental manipulation of membrane lipid composition (Booth et al., 2001; Martens et al., 2000). They would test the possibility that increased input frequency applied to a set of nanowires modeling K_A channels would induce diffusion of unsaturated lipids toward the activated nanowires. In the initial experiment, all compartments would contain only saturated lipids. Activation times and rates of molecular gates would be held constant. Pulses conducted through the membrane should be identical to input. This relationship should persist despite increases in input frequency. The rationale is that saturated lipids, although concentrated in the applied field, do not contain polarized regions. There would thus be no electrostatically-induced nanowire deformation and no interruption of current. This result would model the closed state of the K_A channel correlated with AP propagation success in an actual neuron. In the second set of experiments, activation times and rates of molecular gates would again be held constant, but all compartments would contain an approximately equal mixture of saturated and unsaturated lipids. Input at lower frequencies should produce a stochastic pattern in which pulse lengths approximating input are randomly interspersed with pulses of briefer duration. The rationale is that the more fluid unsaturated lipids would diffuse toward the activated nanowires but their movement would be randomly impeded by complexes of saturated lipids. Unsaturated-lipid concentration, correlated with ethene polarization and nanowire deformation, would thus fluctuate stochastically at lower-frequency input. This predicted transmembrane pattern would model rapid (and random) ion-channel transitions between open and closed states (channel flicker) observed in clusters of voltage-dependent Na+ and K+ channels in cultured hippocampal neurons (Savtchenko et al., 2001b). At higher input frequencies, the dominant nanowire pattern would be a series of brief pulses. The rationale is that the shortened interval between intra-compartmental field effects (<11ms intra-compartmental mean residency time for lipids) would not allow unsaturated lipids to escape into neighboring compartments. This pattern would model the dominant open state of the K_A channel correlated with increased AP failure at higher input frequencies (low-pass filter). In the final experiment, all compartments would contain only unsaturated lipids. Activation times and rates of molecular gates would again be held constant. Input at all frequencies should produce a continuous pattern of brief transmembrane pulses. This pattern would model the gating of a propagating AP in which the prolonged open state of K_A channels is correlated with high probability of propagation failure. Together these protocols would suggest that AP conduction failure is directly correlated with unsaturated-lipid concentration in activated compartments of a K_A channel cluster. Subsequent studies could model K_A channel gating of a propagating AP with increasing realism, and explore its more subtle features. For example, the above protocols could be repeated with changing subsets of targeted nanowires. Here it would be important to examine discrete versus overlapping subsets, and the combined effect of high-frequency input and frequent changes of targeted subsets. Finally, corral gating (held constant in all experiments) could be optically manipulated as a means of inducing diffusion gradients.

5.3. Implications of the Proposed Model

As noted throughout this treatise, for over a half-century neuroscientists have assumed that the Hodgkin-Huxley action potential is the fundamental unit of brain communication. This viewpoint has yielded tremendous insight into the electrophysiological correlates of neural health and disease. Paralleling this history, the development of artificial-intelligence models based on the McCulloch-Pitts neuron has made it possible to generate a wide range of hypotheses regarding normal and abnormal brain states. However, increasing evidence in biophysics and computational modeling suggests that smaller sub-neural systems of corralled lipid microdomains may regulate neural impluse propagation. Thus the neuron may not be a single computational unit, but a set of interacting computational modules. Analogous with the history of the McCulloch-Pitts model, the biophysical and computational data sets signal an opportune moment for developing artificial models of neural membrane molecular systems. While not intended to replace traditional neural-network and artificial-intelligence approaches, supported-membrane models could significantly enrich them by exploring the neuron's computational subtleties. There are several unusual features of neuron electrical activity that might be explored in this way. To select a single example, it is now generally recognized that the strength of the neural impulse can actually increase with greater distance from the soma, contradicting the traditional assumption (in both the Hodgkin-Huxley and McCulloch-Pitts viewpoints) of distance-dependent signal attrition (Magee and Cook, 2000). Spike amplitude results from the duration of the open state of the Na^+ channel during permeant ion movement into the cytosol. In the light of recent evidence that Na^+ channels are distributed in clusters, it is tempting to speculate that the high amplitude of distal spikes could result from channel-cluster enrichment with a high concentration of unsaturated lipids (Rasband et al., 1999). In terms of the present model, this feature would increase the complexity of membrane/Na^+ channel electrostatic interactions, thus prolonging the channel open time, and increasing the spike amplitude. A plausible functional result of high-amplitude distal spikes is greater likelihood of transmission across the synapse, thus increasing the activity in the network, and counteracting to some extent the effect of K_A channel clusters. More sophisticated artificial models could address these interactions. It is now virtually certain that the human brain uses more than one type of coding. An increased understanding of brain information-processing on the molecular level may enrich our understanding of both normal and pathological function.

Chapter VI

Conclusion

Initially ignored by early cell biologists because of the instrumental limitations of their day, and subsequently dismissed as a structure for anchoring proteins, the cell membrane is assuming its rightful place as a focus of biological research. Although the precise mechanisms by which the membrane regulates changes in protein conformation integral to cellular processes are the object of current controversy, there remains little serious doubt that the membrane has significant cell-regulatory functions. This small book has examined the possibility that electrostatic interactions between polarized ethenes in neural membrane lipids and charge residues in a gated ion channel comprise a molecular mechanism for regulating neural communication. If subsequent experiments demonstrate that this model (or something similar to it) is an accurate portrayal of nature, there would be important implications for artificial intelligence and medicine.

At a minimum, the traditional Mc-Culloch-Pitts model (and the models it spawned) which views the neuron as the fundamental element of brain information-processing, would undergo significant revision. New architectures exploring the systemic implications of such phenomena as neural impulse gating, and progressive increase in spike amplitude at increasing distance from the soma exemplify the possible research directions resulting from the change in viewpoint. Some of these novel architectures could have medical applications. One possibility, explored in this book, is the use of a supported-membrane architecture to examine fundamental properties of the A-current potassium channel, implicated in epilepsy. There is no compelling reason for believing that other neural diseases, with identified molecular etiologies, could not also be explored in this way. Additionally, as this architecture becomes better understood, perhaps it could embedded into the model neurons of traditional artificial-intelligence approaches, thus linking the molecular level with the larger systemic level. More direct applications to molecular medicine, in the form of drug-delivery systems could also derive from the model. Although important obstacles such as drug movement past the blood-brain barrier would have to be overcome, it seems possible that delivery systems based on the regulatory properties of lipids could become a feature of new neural therapies. Perhaps a vesicle tagged with two molecules, one to breach the blood-brain barrier, the other to target an ion-channel region, could be used to therapeutically modify the ion channel's molecular environment.

On a final note I should also point out that, as an anthropologist, I am intrigued by the model's possible implications for the study of molecular neural evolution. As I noted in the Introduction, most of the studies done in this area have emphasized ion channels. However (reflecting an earlier scientific bias, now largely past) comparatively few evolutionary studies have been done on the neural membrane. Advances in living-membrane and artificial-membrane studies described in earlier chapters, combined with a now extensive data set in comparative neurobiology, suggest an opportune moment to pursue these types of studies. As a single example, one might explore the adaptive advantage of highly optimized molecular controls over conduction failure at branch points in cortico-cortical networks. Communication between and within cortical regions, especially association areas, is obviously critical for species survival. However, the branching geometry that makes this communication possible invites conduction failure because of impedance mismatch. In mammalian species, are there highly optimized molecular systems for overcoming mismatch effects and insuring reliable impulse conduction? For example, does one find, at branch points, clusters of Na^+ channels surrounded by microdomains enriched with unsaturated lipids? In principle, this structure would enhance conduction success by prolonging the open time of the channels, and achieving reversal of membrane potential above the local depolarization threshold. Additionally, it seems likely that a molecular architecture of this type would require sequestration from changing membrane composition. Accordingly, the cytoskeleton fence surrounding the channel cluster/microdomain complex would probably be equipped with fast spectrin gates, or arguably with no gates at all. (This possibility, in turn, invites the question of plasticity versus hard-wiring in the brain's computational systems having its counterpart on the molecular level.) In any event, it is clear that neural membrane studies will comprise an exciting dimension of 21st century neuroscience. Over the generations, a deeper level of order has been discovered: The unfolding of its scientific and medical implications awaits us.

Chapter VII

Appendix: Table and Figures

Table 1. Claculated moments and polarizabilities of ethene model compounds*

Dipole moment (Debye)				
Ethene	X = 0.0000	Y = 0.0000	Z = 0.0000	Total = 0.0000
Ethene dimer	X = -0.0001	Y = -0.5306	Z = 0.0794	Total = 0.5365
Ethene trimer	X = 0.0005	Y = -0.6531	Z = -0.0306	Total = 0.6538
Quadrupole moment (Debye -Å)				
Ethene	XX = -15.1785	YY = -12.1966	ZZ = -11.9487	
Ethene dimer	XX = -48.4728	YY = -43.6183	ZZ = -42.6662	
Ethene trimer	XX = -75.6495	YY = -67.8488	ZZ = -67.3202	
Polarizability ($J^{-1}C^2m^2$)				
Ethene	α_{xx} = 7.672	α_{yy} = 19.226	α_{zz} = 30.875	$<\alpha>$ = 19.258
Ethene dimer	α_{xx} = 54.029	α_{yy} = 84.076	α_{zz} = 62.289	$<\alpha>$ = 66.798
Ethene trimer	α_{xx} = 93.644	α_{yy} = 125.284	α_{zz} = 91.345	$<\alpha>$ = 103.424

Geometrics of model compounds were optimized using the 4-3JG basis set. Polarizabilities were obtained by performing frequency calculations on geometry using the same basis set. The mean polarizability, $<\acute{\alpha}> = (\acute{\alpha}_{xx} + \acute{\alpha}_{yy} + \acute{\alpha}_{zz})/3$.

Table 2. Electric moments of model compounds[a, b, c]

Moment	*Type of ethene*			
	Monomer	*Dimer (stacked)*	*Dimer (extended)*	*Trimer (stacked)*
Dipole (μ_{total})	0.000	0.536	0.095	0.654
Quadrupole ($\theta_{average}$)	-13.162 (**1.0**)	-24.629 (**1.9**)	-31.283 (**2.4**)	-37.299 (**2.8**)
Polarizability ($\alpha_{average}$)	19.258 (**1.0**)	37.357 (**1.9**)	40.026 (**2.1**)	56.508 (**2.9**)

[a]Units of dipole moment are Debye, Units of quadropole moment are Debye Á. Units of polarizability are Bohr^3. 1 Bohr = 0.529Á.

[b]Quadropole moments and polarizabilities have been corrected for contributions of linker atoms, $\theta_{average}$ and $\acute{\alpha}_{average}$ represent the spatially averaged quantity which is determined by adding the xx, yy, and zz components and dividing by 3.

[c]Values shown in bold in parenthesis represent a ratio with respect to an ethane monomer.

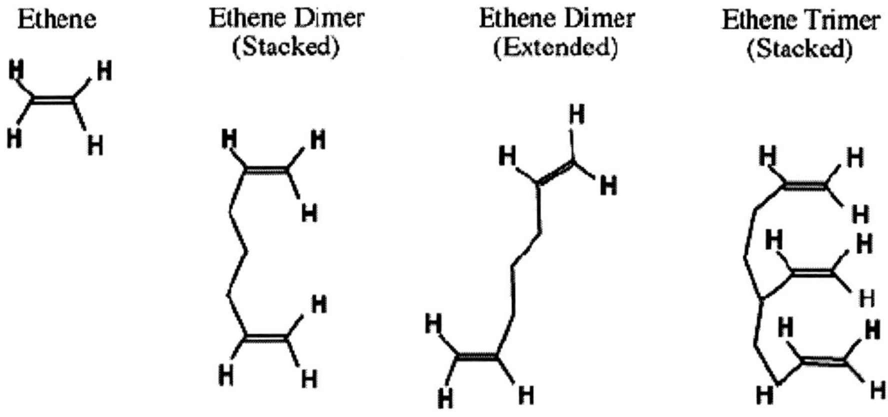

Figure 1. Ethene model compounds. Stacked and extended models represent spatial distributions of ethenes within organized and randomized microdomains, respectively.

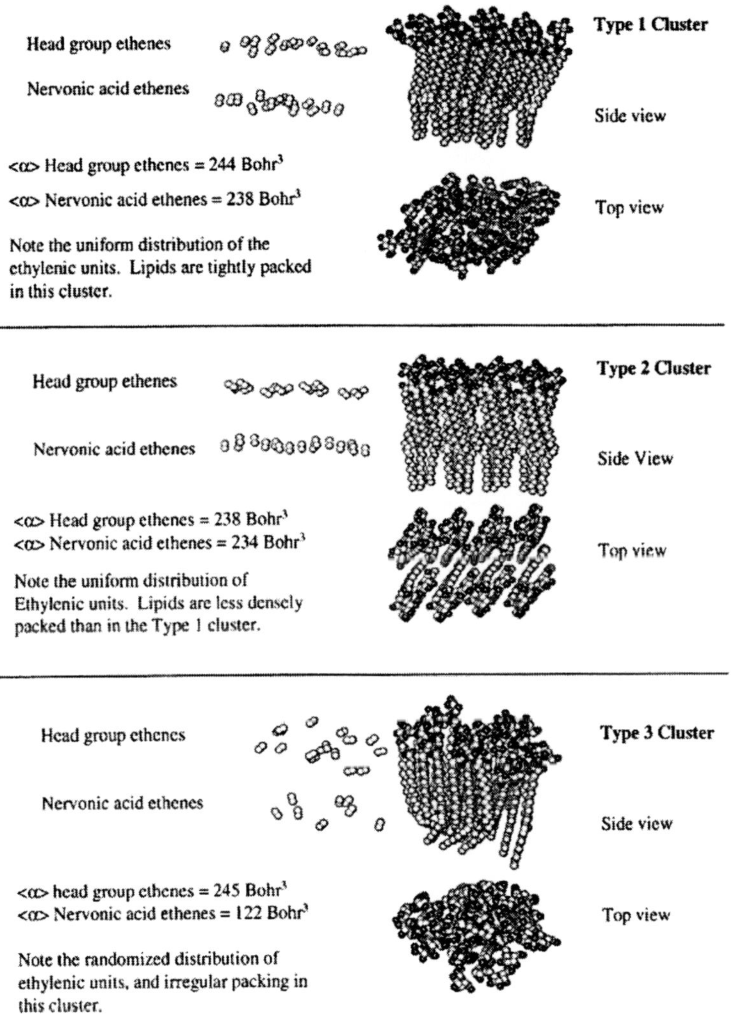

Figure 2. Energetically favorable galactocerebroside clusters. Type 1 and 2 clusters are composed of 16 nervonic acid-containing galactocerebroside molecules. The Type 3 cluster is composed of 8 steric acid containing galactocerebroside and 8 nervonic containing galactocerebroside. Space-filing models of these lipids are provided in Figure 1. The average polarizability, ά of ethenes in the head group and nervonic acid chain for each cluster is included. This property links the field-induced pertutbaton of the charge distribution of an ethylenic bond to its ability to interact with the field.

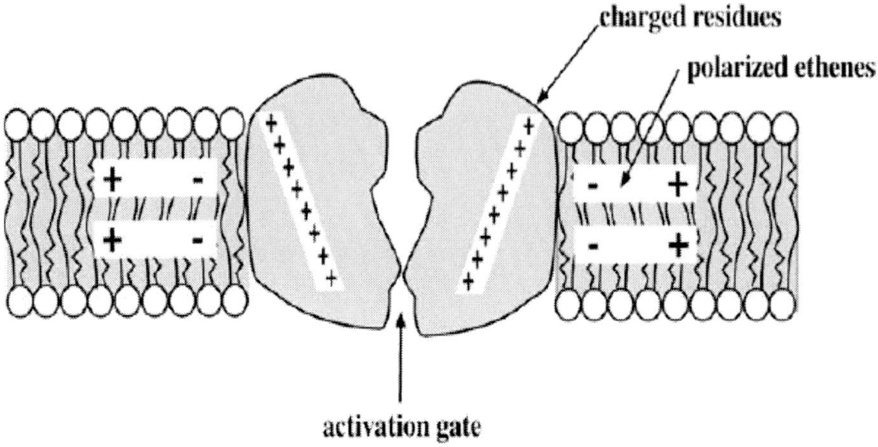

Figure 3. Model of membrane microdomain regulation of $K_A$channel activity at branch points. Polarized ethenes in membrane interact wiuth charged residues in the gated channel. Ethene charge localization is variable.

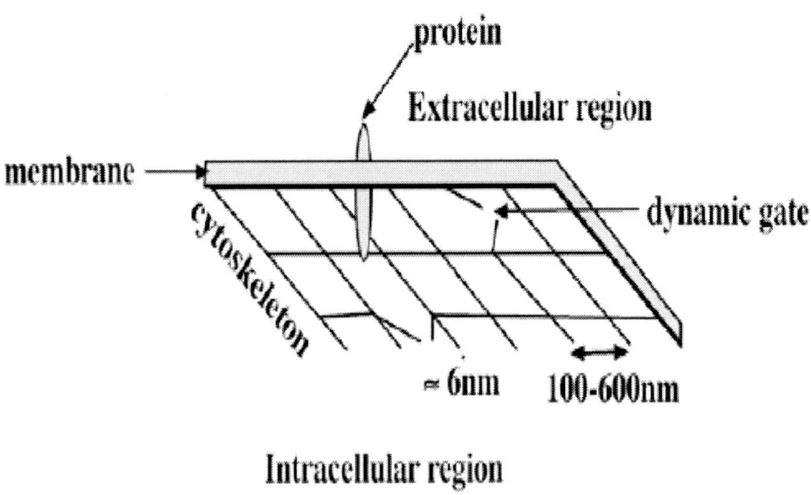

Figure 4. Schematic diagram of cytoskeleton matrix. A thickness of 6nm and compartment size of 100-600nm are indicated.Dynamic gating results from stochastic dissociation and reassociation of spectrim tetramers. Gating permits stepwise changes in relative concentration of saturated and unsaturated microdomains within a corral. Adapted from Ref [49].

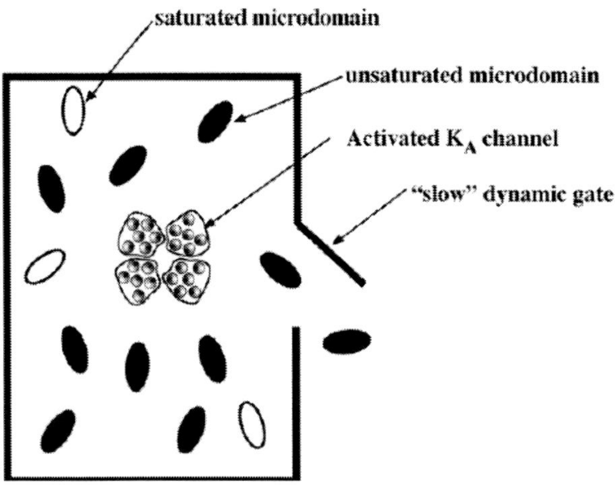

Figure 5. Molecular structure consistent with AP conduction failure at branch points. High concentration of unsaturated microdomains prolongs K_A channel "open" time. Slow cytoskeleton gate permits field-induced hop diffusion of unsaturated microdomains from neighboring corrals.

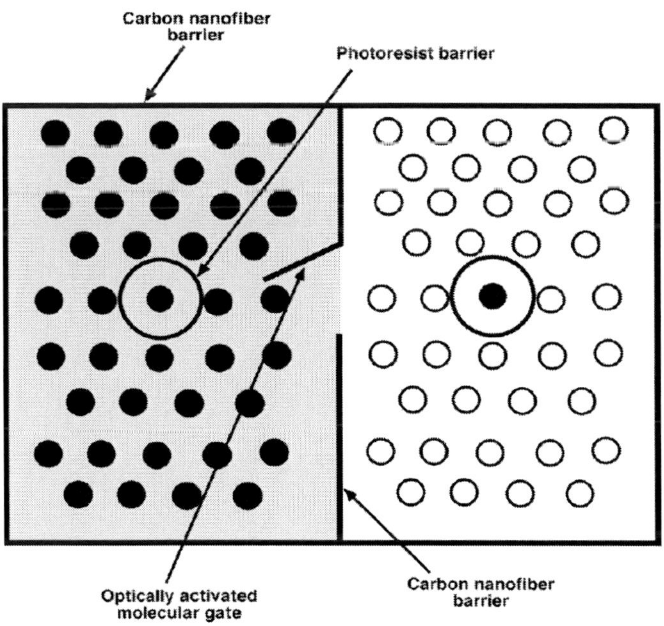

Figure 6. Supported membrane (topview) with activated biotemplated nanowire. Nanowire field has produced a phase transition between saturated (O) and unsaturated (•) lipids. For effect of unsaturated lipids on activated nanowire see Figure 7.

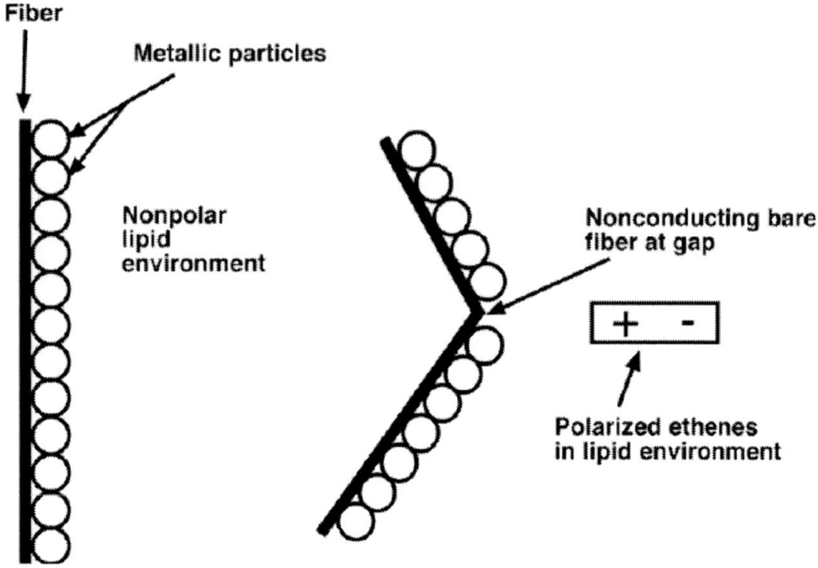

Figure 7. Activation of metallized nanowire polarizes ethenes in lipid environment (side view). Electrostatic interaction between polarized ethenes and ananowire produces transient nanowire deformation and nonconducting state.

References

Alfsen, A., 1989. Membrane dynamics and molecular traffic and sorting in mammalian cells. *Prog. Biophys. Mol. Biol.* 54, 145-157.

Antonucci, D., Lim, S., Vassanelli, S., Trimmer, J., 2001. Dynamic localization and clustering of dendritic Kv2.1 voltage-dependent potassium channels in developing hippocampal neurons. *Neuroscience* 108, 69-81.

Atmar, W., 1994. Notes on the simulation of evolution. *IEEE Trans. Neural. Netw.* 5, 130-147.

Baroni, A., Paoletti, I., Silvestri, I., Buommino, E., Carriero, M., 2003. Early vitronectin downregulation in a melanoma cell line during all-trans retinoic acid-induced apoptosis. *Br. J.Dermatol.* 148(3), 424-433.

Barron, D., Matthews, B., 1935. Intermittent conduction in the spinal cord. *J. Physiol.* 85, 73-103.

Birnbaum, S., Varga, A., Yuan, L.-L., Anderson, A., Sweatt, J., Schrader, L., 2004. Structure and function of Kv4-family transient potassium channels. *Physiol. Rev.* 84, 803-833.

Biron, E., Otis, F., Meillon, J.-C., Robitaille, M., Lamothe, J., Van Hove, P., Cormier, M.-E., Voyer, N., 2004. Design, synthesis and characterization of peptide nanostructures having ion-channel activity. *Bioorg. Med. Chem.* 12, 1279-1290.

Bogoslovskaya, I., Lyubinskii, I., Pozin, N., Putsillo, Y., Shmelev, I., Shura-Bura, T., Spread of excitation along a fiber with local inhomogeneities. *Biophysics* 18, 944-948.

Boiko, T., Winckler, B., 2003. Picket and other fences in biological membranes. *Dev. Cell.* 5, 191-192.

Booth, P., Templer, R., Curran, A., Allen, S., 2001. Can we identify the forces that drive the folding of integral membrane proteins? *Biochem. Soc. Trans.* 29, 408-413.

Booth, P., Curran, A., 1999. Membrane protein folding. *Curr. Opin. Struc. Biol.* 9, 115-121.

Branton, D., Park, R. 1968a. Introduction. In: Branton, D., Park, R., (Eds.), *1968b Papers on Biological Membrane Structure.* Little, Brown, and Company, Boston.

Branton, D., Park, R., (Eds.), 1968b. *Papers on Biological Membrane Structure.* Little, Brown and Company, Boston.

Brown, R., 1998. Sphingolipid organization in biomembranes: what physical studies of model membranes reveal. *J. Cell Science* 111, 1-9.

Brown, F., Leitner, D., McCammon J., Wilson K., 2000. Lateral diffusion of membrane proteins in the presence of static and dynamic corral: suggestions for appropriate observables. *Biophys J.*78, 2257-2269.

Buldum, A., Lu, J., 2003. Electron field emission properties of closed and open carbon nanotubes. In: *Technical Proceedings of the 2003 Nanotechnology Conference and Trade Show,* vol.3, Nanotech 2003, San Francisco, pp. 297-300.

Cantor, R., 1999. Lipid composition and the lateral pressure profile in bilayers. *Biophys. J.* 76, 2625-2639.

Catterall, W., 2002. Molecular mechanisms of gating and drug block of sodium channels. *Novartis Found. Symp.* 241, 206-218.

Churchland, P., Sejnowski, T., 1992. *The Computational Brain.* The MIT Press, Cambridge.

Cooper, G., 1997. *The Cell: A Molecular Approach.* Sinauer Associates, Sunderland.

Cornea, R., Thomas, D., 1994. Effects of membrane thickness on the molecular dynamics and enzymatic activity of reconstituted Ca-ATPase. *Biochemistry* 33(10), 2912-2920.

Cragg, P., 2002. Artificial transmembrane channels for sodium and potassium. *Sci. Progr.* 85, 219-241.

Dai, H., 2002. Carbon nanotubes: synthesis, integration, and properties. *Acc. Chem. Res.* 35, 1035-1044.

Danielli, J. and Davson, H., 1935. A contribution to the theory of permeability of thin films. *J. Cell Comp. Physiol.*5: 495-508.

Dawkins, R., 1986. *The Blind Watchmaker: Why the Evidence of Evolution Reveals a Universe Without Design.* W.W. Norton and Company, New York.

Dawkins, R. 1997. *Climbing Mount Improbable.* W.W. Norton and Company, New York.

Debanne, D., Kopysova, I., Bras, H., Ferrand, N., 1999. Gating of action potential propagation by an axonal A-like potassium conductance in the hippocampus: a new type of non-synaptic plasticity. *J. Physiol.* Paris 93, 285.

Delgado-Escueta, A., Wilson, W., Olsen, R., Porter, R., 1999. New waves of research in the epilepsies: Crossing into the third millennium. In: Delgado-Escueta, A., Wilson, W., Olsen, R., Porter, R., (Eds.), *Jasper's Basic Mechanisms of the Epilepsies.* Vol. 79. Lippincott Williams and Wilkins, Philadelphia.

De Planque, M., Goormaghtigh, E., Greathouse, D., Koeppe, R. 2nd, Kruijtzer, J., Liskamp, R., de Kruijff, B., Killian, J., 2001. Sensitivity of single membrane-spanning alpha-helical peptides to hydrophobic mismatch with a lipid bilayer: effects on backbone structure, orientation, and extent of membrane incorporation. *Biochemistry* 40(16), 5000-5010.

Destexhe, A., Rudolph, M., Pare, D., 2003. The high conductance state of neocortical neurons in vivo. *Nature Rev. Neurosci.* 4 (9), 739-751.

Dichter, M., 1994. Emerging insights into mechanisms of epilepsy: Implications for new antiepileptic drug development. *Epilepsia* 35, 551-557.

Dichter, M., 1997. Basic mechanisms of epilepsy: Targets for therapeutic intervention. *Epilepsia* 38, 52-56.

Doktycz, M., Zhang, L., Melechko, A., Klein, K., McKnight, T., Britt, P., Guillorn,M., Merkulov, V., Lowndes, D., Simpson, M., 2003. Nanofiber structures as mimics for cellular membranes. *Nanotechnology*, 420-423.

Dresselhaus, M., Dresselhaus, G., Ecklund, P., 1996. *Science of Fullerenes and Carbon Nanotubes.* Academic Press, New York.

Drexler, K., 1994. Molecular nanomachines: Physical principles and implementation strategies. In: Stroud, R., Cantor, C., and Pollard, T. (Eds.), *Annual Review of Biophysics and Biomolecular Structure.* Annual Reviews Inc., Palo Alto.

Dumas, F., Lebrun, M., Tocanne, J-F., 1999. Is the protein/lipid hydrophobic matching principle relevant to membrane organization and functions? *FEBS Letters* 458, 271-277.

Dumas F., Sperotto M., Lebrun M., Tocanne J., Mouritsen O., 1997. Molecular sorting of lipids by bacteriorhodopsin in dilauroylphosphatidylcholine/distearoylphosphatidylcho line lipid bilayers. *Biophys. J.* 73, 1940-1953.

Dykstra, C., 1997. *Physical Chemistry: A Modern Introduction*. Prentice Hall, Upper Saddle River.

Edidin, M., 1989. Fluorescent labeling of cell surfaces. In: Wang, Y-L., Taylor, D., (Eds.), *Fluorescent microscopy of living cells in culture: Part A*. Academic Press, New York.

Emilien, G., Maloteaux, M., 1998. Pharmacological management of epilepsy. Mechanism of action, pharmacokinetic drug interactions, and new discovery possibilities. *Int. J. Clin. Pharmacol. Ther.* 36, 181-194.

Fleshman, J., Segev, I., Burke, R., 1988. Electrotonic architecture of type identified α-motoneurons in the cat spinal cord. *J. Neurophysiol.* 60, 60-85.

Frisch, M., Trucks, G., Schlegel, H., Gill, P., Johnson, B., Robb, M., Cheeseman, J. Keith, T., Petersson, G., Montgomery, J., Raghavachari, K., Al-Laham, M., Zakrzewski, V., Ortiz, J., Foresman, J., Cioslowski, J., Stefanov, B., Nanayakkara, A., Challacombe, M., Peng, C., Ayala, P., Chen, W., Wong, M., Andres, J., Replogle, E., Gomperts, R., Martin, R., Fox, D., Brinkley, J., Defrees, D., Baker, J., Stewart, J., Head-Gordon, M., Gonzalez, C., Pople, J., 1995. Gaussian Inc., Pittsburgh.

Feringa, B., Komura, N., Van Delden, R., Ter Wiel, M., 2002. Light-driven molecular switches and rotors. *Appl. Phys*. A 75, 301-308.

Frye, L., Edidin, M., 1970. The rapid intermixing of cell surface antigens after formation of mouse-human heterokaryons. *J. Cell Sci.* 7, 319-335.

Fujiwara, T., Ritchie, K., Murakoshi, H., Jacobson, K., Kusumi, A., 2002. Phospholipids undergo hop diffusion in compartmentalized cell membrane. *J. Cell Biol.* 157, 1071-1081.

Garey, M., Johnson, D., 1979. *Computers and Intractability: A Guide to the Theory of NP-Completeness*. Freeman, San Francisco.

Geiger, B., Ayalon, O., Ginsberg, D., Volberg, T., Rodriguez Fernandez, J., Yarden, Y., Ben-Ze'ev, A., 1992. Cytoplasmic control of cell adhesion. *Cold Spring Harb. Symp. Quant.* Biol. 57, 631-642.

Gennis, R., 1989. Biomembranes: Molecular Structure and Function. *Springer*, New York.

Gokel, G., Schlesinger, P., Djedovič, N., Ferdani, R., Harder, E., Hu, J., Leevy, W., Pajewska, J., Weber, M., 2004. Functional, synthetic organic chemical models of cellular ion channels. *Bioorg. Med. Chem.* 12, 1291-1304.

Goldstein, S., Rall, W., 1974. Changes in action potential shape and velocity for changing core conductor geometry. *Biophys. J.* 14, 731-757.

Gorter, E., Grendel, F., 1968 (orig. 1925). On bimolecular layers of lipoids on the Chromocytes of the blood. . In: Branton, D., Park, R., (Eds.), 1968. *Papers on Biological Membrane Structure.* Little, Brown and Company, Boston.

Gray, J., 1982. Precis of the neuropsychology of anxiety: an enquiry into the functions of the septo-hippocampal system. *Behavioral and Brain Sciences* 5, 469-484.

Green, D. (Ed.), 1972. Membrane structure and its biological applications. *Annals of The New York Academy of Sciences*, New York.

Grossi, E., Buscema, M., Snowdon, D., Antuono, P., 2007. Neuropathological findings processed by artificial neural networks (ANNs) can perfectly distinguish Alzheimer's patients from controls in the Nun Study. *BMC Neurol.* 7:15, http://www.biomedcentral.com

Groves, J., Boxer, S., McConnell, H., 1997. Electric field-induced reorganization of two-component supported bilayer membranes. *Proc. Natl. Acad. Sci. USA* 94, 13990-13995.

Groves, J., Ulman, N., Cremer, P., Boxer, S., 1998. Supported-membrane interactions: Mechanisms for imposing patterns on a fluid lipid bilayer membrane. *Langmuir* 14, 3347-3350.

Hall, Z. (Ed.), 1992. *An introduction to molecular neurobiology.* Sinauer Associates. Sunderland.

Hao, M., Mukherjee, S., Maxfield, F., 2001. Cholesterol depletion induces large-scale domain segregation in living cell membranes. *Proc. Natl. Acad. Sci. USA* 98 (23), 13072-13077.

Harris, H., 1999. *The Birth of the Cell.* Yale University Press, New Haven.

Harris-Warwick, R., 2000. Ion channels and receptors: Molecular targets for behavioral evolution. *J. Comp Physiol.* [A];186, 605-16.

Hatt, H., Smith, D., 1975. Axon conduction block: differential channeling of nerve impulses in the crayfish. *Brain Res.* 87, 85-88.

Hering, H., Lin, C-C., Sheng, M., 2001. Lipid rafts in the maintenance of synapses, dendritic spines, and surface AMPA receptor stability. *J. Neurosci.*23, 3262-3271.

Hille, B., 2001. *Ionic Channels of Excitable Membranes.* Sinauer Associates, Sunderland.

Hodgkin, A., Huxley A., 1952a. The components of membrane conductance in the giant axon of *Loligo*. *J. Physiol.* 116, 473-496.

Hodgkin, A., Huxley A., 1952b. Currents carried by sodium and potassium ions Through the membrane of the giant axon of *Loligo*. *J. Physiol.* 116, 449-472.

Hodgkin, A., Huxley A., 1952c. The dual effect of membrane potential on sodium Conductance in the giant axon of *Loligo*. *J. Physiol.* 116, 497-506.

Hodgkin, A., Huxley A., 1952d. A quantitative description of membrane current and its application to conduction and excitation in nerve. *J. Physiol.* 117, 500-544.

Hong, K, Miller, C., 2000. The lipid-protein interface of a *Shaker* K^+ channel. *J. Gen. Physiol.* 115, 51-58.

Hughes, A., 1989. A History of Cytology. Iowa State University Press, Ames.

Jain, M., 1972. *The Bimolecular Lipid Membrane: A System*. Van Nostrand Reinhold, New York.

Jefferys, J., 1993. The pathophysiology of epilepsies. In: Laidlaw, J, Richens, A., Chadwick, D., (Eds.), *A Textbook of Epilepsy*. Churchill and Livingstone, New York.

Jørgensen, K., Mouritsen, O., 1995. Phase separation dynamics and lateral organization of two-component lipid membranes. *Biophys. J.* 95, 942-954.

Kanicky, J., Shah, D. n.d. Effect of degree, type, and position of unsaturation on the pKa of long-chain fatty acids. Unpublished manuscript. *Center for Surface Science and Engineering*. University of Florida.

Killian, J., 1998. Hydrophobic mismatch between proteins and lipids in membranes. *Biochim. Biophys. Acta* 1376, 401-416.

Kinnunen, P., 1991. On the principles of functional ordering in biological membranes. *Chem. Phys. Lipids* 57, 375-399.

Kinnunen, P., Virtanen, J., 1986. A qualitative, molecular model of the nerve impulse: conductive properties of unsaturated lyotropic liquid crystals. In: Gutman, F. and Keyzer, H., (Eds.), *Modern Bioelectrochemistry*. Plenum Press, New York.

Kirkpatrick, L., Brady, S., 1999. Cytoskeleton of neurons and glia. In: Siegel, G. (Ed.), *Basic Neurochemistry*. Lippincott, Williams, and Wilkins, Philadelphia.

Koch, C., 1999. *Biophysics of computation: information processing in single neurons*. Oxford University Press, New York.

Koch, C., Segev, I., 2000. The role of single neurons in information processing. *Nat. Neurosci. Suppl.* 3, 1171-1177.

Kopysova, I., Debanne, D., 1998. Critical role of axonal A-type K^+ channels and axonal geometry in the gating of action potential propagation along CA3 pyramidal cell neurons: a simulation study. *J. Neurosci.* 18, 7436-7451.

Krnjevic, K., Miledi, R., 1959. Presynaptic failure of neuromuscular propagation in rats. *J. Physiol.* 149, 1-22.

Kusumi, A., Nakada, C., Ritchie, K., Murase, K., Suzuki, K., Murakoshi, H., Kasai, R. Kondo, J., Fujiwara, T., 2005. Paradigm shift of the plasma membrane concept from the two-dimensional continuum fluid to the partitioned fluid: high-speed single-molecule tracking of membrane molecules. *Annu. Rev. Biophys. Biomol. Struct.* 34, 351-378.

Langmuir, I., 1917. The constitution and fundamental properties of solids and liquids. II. Liquids. *J. Am. Chem. Soc.* 39, 1848-1906.

Lauritzen, I., Blondeau, N., Heurteaux, C., Widmann, C., Romney, G., Lazdunski, M., Polyunsaturated fatty acids are potent neuroprotectors. *EMBO J.* 19, 1784-1793.

Le Coutre, J., Narasimhan, L., Kumar N. Patel, C., Kaback, R., 1997. The lipid bilayer determines helical tilt angle and function in lactose permease of *Escherichia coli*. *Proc. Natl. Acad. Sci. USA.* 94, 10167-10171.

Lee, K., Klingler, J., McConnell, H., 1994. Electric-field-induced concentration Gradients in lipid monlayers. *Science* 263, 655-658.

Lehtonen, J., Holopainen, J., Kinnunen, P., 1996. Evidence for the formation of microdomains in liquid crystalline large unilamellar vesicles caused by hydrophobic mismatch of the constituent phospholipids. *Biophys. J.* 70, 1753-1760.

Lehtonen, J., Kinnunen, P., 1997. Evidence for phospholipid microdomain formation in liquid crystalline liposomes reconstituted with *Escherichia coli* lactose permease. *Biophys. J.* 72, 1247-1257.

Levitan, I., Kaczmarek, L., 2002. *The Neuron: Cell and Molecular Biology*. Oxford University Press, New York.

Leitner, D., Brown, F., Wilson, K., 2000. Regulation of protein mobility in cell membranes: a dynamic corral model. *Biophys. J.* 78, 125-135.

Lewis, B., Engelman, D., 1983. Bacteriorhodopsin remains dispersed in fluid phospholipid bilayers over a wide range of bilayer thicknesses. *J. Mol. Biol.* 166(2), 203-210.

Li, X-M, Momsen, M., Smaby, J., Brockman, H., Brown, R., 2001. Cholesterol decreases the interfacial elasticity and detergent solubility of sphingomyelins. *Biochemistry* 40, 5954-5963.

Liang, J., 2002. Experimental and computational studies of determinants of membrane-protein folding. *Curr. Opin. Chem. Biol.* 6, 878-884.

Llinas, R., Nicholson, C., and Precht, W., 1969. Preferred centripetal conduction of dendritic spikes in alligator Purkinje cells. *Science* 163, 184-187.

Madeja, M., 2000. Extracellular surface charges in voltage-gated ion channels. *News Physiol. Sci.* 15, 15-19.

Magee, J., Cook, E., 2000. Somatic EPSP amplitude is independent of synapse location in hippocampal pyramidal neurons. *Nature Neurosci.* 3, 895-903.

Maiti, A., Brabec, C., Roland, C., Yakobson, B., Bernhole, J., 1996. Growth and elastic properties of nanotubes. In: *Eighth Annual Workshop on Recent Developments in Electronic Structure Algorithms*. Minneapolis, MN.

Marsh, D., 1995. Lipid-protein interactions and heterogeneous lipid distribution in membranes. *Mol. Membrane Biol.* 12, 59-64.

Martens, J., Polanco-Navarro, R., Coppock, E., Nishiyama, A., Parshley, I., Grobaski, T., Tamkun, M., 2000. Differential targeting of Shaker-like potassium channels to lipid rafts. *J. Biol. Chem.* 275, 7443-7446.

Martens, J., O'Connell, K., Tamkun, M., 2004. Targeting of ion channels to membrane microdomains: localization of Kv channels in lipid rafts. *Trends Pharmacol. Sci.* 25, 16-21.

McConnell, H., Radhakrishnan, A., 2003. Condensed complexes of cholesterol and Phospholipids. *Biochim. Biophys. Acta* 1610, 159-173.

McCullough, W., Pitts, W., 1943. A logical calculus of the ideas immanent in nervous activity. *Bull. Math. Biophys.* 5, 115-133.

McNamara, J., 1999. Emerging insights into the genesis of epilepsy. Nature 24, A15-22.

Mel, B., 2002. What the synapse tells the neuron. *Science* 295, 1845-1846.

Mitchell, D., Litman, B., 1998. Effect of cholesterol on molecular order and dynamics in highly polyunsaturated phospholipid bilayers. *Biophys. J.* 75, 896-908.

Mobashery, N., Nielsen, C., Andersen, O., 1997. The conformational preference of gramicidin channels is a function of lipid bilayer thickness. *FEBS Letters* 412, 15-20.

Mouritsen, O., Bloom, M.., 1984. Mattress model of lipid-protein interaction in membranes. *Biophys. J.* 46(2), 141-153.

Mouritsen, O., Bloom, M., 1993. Models of lipid-protein interactions in membranes. In: Engleman, D., Cantor, C., Pollard, T. (Eds.), Annual Review of Biophysics and Biomolecular Structure, Vol. 22. *Annual Reviews*, Palo Alto.

Mouritsen, O., Jørgensen, K., 1992. Dynamic lipid bilayer hetereogeneity: a mesoscopic vehicle for membrane function? *BioEssays* 1, 129-136.

Mouritsen, O., Jørgensen, K., 1994. Dynamical order and disorder in lipid bilayers. *Chem. Phys. Lipids* 73, 3.

Mouritsen, O., Kinnunen, P., 1996. Role of lipid organization and dynamics for membrane functionality. In: Merz, K., Roux, B. (Eds.), *Biological membranes: A molecular perspective from computation and experiment.* Birkhäuser, Boston.

Nakada, C., Ritchie, K., Oba, Y., Nakamura, M., Hotta, Y., Iino, R., Kasai, R., Yamaguchi, K., Fujiwara, T., Kusumi, A., 2003. Accumulation of anchored proteins forms membrane diffusion barriers during neuronal polarization. *Nat. Cell Biol.* 5, 626-632.

Nelson, W., Wilson, R., Wollner, D., Mays, R., McNeill, H., Siemers, K., 1992. Regulation of epithelial cell polarity: a view from the cell surface. *Cold Spring Harb. Symp. Quant. Biol.* 57, 621-630.

Nestler, E., Hyman, S., Malenka, R., 2001. *Molecular Neuropharmacology: A Foundation for Clinical Neuroscience*. McGraw-Hill, New York.

Overton, E., 1968 (Orig., 1899). The probable origin and physiological significance of cellular osmotic properties. In: Branton, D., Park, R., (Eds.), 1968. *Papers on Biological Membrane Structure*. Little, Brown and Company, Boston.

Papazian, D., 1999. Potassium channels: some assembly required. *Neuron* 23, 7-10.

Papazian, D., Shao, X., Seoh, S.-A., Mock, A., Huang, Y., Wainstock, D., 1995. Electrostatic interactions of S4 voltage sensor in Shaker K^+ channel. *Neuron* 14, 1293-1301.

Parnas, I., 1972. Differential block at high frequencies of branches of a single axon innervating two muscles. *J. Neurophysiol.* 35, 903-914.

Peschke, J., Riegler, J., Möhwald, H., 1987. Quantitative analysis of membrane distortions introduced by mismatch of protein and lipid hydrophobic thickness. *Eur. Biophys*. J. 14, 385-391.

Petersen N., Hoddelius P., Wiseman P., Seger O., Magnusson K., 1993. Quantification f membrane receptor distributions by image correlation spectroscopy: concept and application. *Biophys. J.* 65, 1135-1146.

Pink, D., Chapman, D., 1979. Protein-lipid interactions in bilayer membranes: a lattice model. *Proc. Natl. Acad. Sci. USA* 76(4), 1542-1546.

Poo, M., Robinson, K., 1977. Electrophoresis of concanavalin A receptors along embryonic muscle cell membrane. *Nature* 17, 602-605.

Price, H., Wallace, R., 2001. A computational model of membrane lipid electronic properties in relation to neural signaling. *BioSystems* 59, 27-34.

Putney, J., 2003. Capacitative calcium entry in the nervous system. *Cell Calcium* 34, 339-344.

Radhakrishnan, A., McConnell, H., 2000. Electric field effect on cholesterol-phospholipid complexes. *Proc. Natl. Acad. Sci. USA* 97, 1073-1078.

Rall, W., 1977. Dendritic spines and synaptic potency. In: Porter, R., (Ed.), *Studies in neurophysiology*. Cambridge University Press, New York.

Rall, W., Burke, R., Holmes, W., Jack, J., Redman, S., Segev, I., 1992. Matching dendritic neuron models to experimental data. *Physiol. Rev.* 72, S159-S186.

Rasband, M., Schrager, P., 2000. Ion channel sequestration in central nervous system axons. *J. Physiol.* 525, 63-73.

Robertson, J., 1957. New observations on the ultrastructure of the membranes of frog peripheral nerve fibers. *J. Biophys. Biochem. Cytol.* 3, 1043-1047.

Rochefort, A., Avouris, P., Lesage, F., Salahub, D., 1999. Electrical and mechanical properties of distorted carbon nanotubes. *Phys. Rev.* B. 60, 13824-13830.

Sadava, D., 1993. Cell Biology: Organelle Structure and Function. Jones and Bartlett, Boston.

Salaun, C., James, D., Chamberlain, L., 2004. Lipid rafts and the regulation of exocytosis. *Traffic* 5, 255-264.

Samsonov, A., Mihalyov, I., Cohen, F., 2001. Characterization of cholesterol-sphingomyelin domains and their dynamics in bilayer membranes. *Biophys. J.* 81, 1486-1500.

Sargent, P., 1992. Electrical Signaling. In: Hall, Z. (Ed.), *An Introduction to Molecular Neurobiology*. Sinauer Associates, Sunderland.

Savtchenko, L., Gogan, P., Korogod, S., Tyč-Dumont, S., 2001a. Imaging stochastic spatial variability of active channel clusters during excitation of single neurons. *Neurosci. Res.* 39, 431-446.

Savtchenko, L., Gogan, P., Tyč-Dumont, S., 2001b. Dendritic spatial flicker of local membrane potential due to channel noise and problematic firing of hippocampal neurons in culture. *Neurosci. Res.* 41, 161-183.

Scheibel, T., Parthasarathy, R., Sawicki, G., Lin, X.-M., Jaeger, H., Lindquist, S., 2003. Conducting nanowires built by controlled self-assembly of amyloid fibers and Selective metal deposition. *Proc. Natl. Acad. Sci.* 100, 4527-4532.

Scott, A., 1995. *Stairway to the Mind: The Controversial New Science of Consciousness*. Copernicus, New York.

Shin, C., McNamara, J., 1994. Mechanisms of epilepsy. *Annu. Rev. Med.* 45, 379-389.

Simons, K., Dupree, P., Fiedler, K., Huber, L., Kabayashi, T., Kurzchalia, T., Olkkonen, V., Pimplikar, S., Parton, R., Dotti, C., 1992. Biogenesis of cell-surface polarity in epithelial cells and neurons. *Cold Spring Harb. Symp. Quant. Biol.* 57, 611-619.

Simons, K., Ikonen, E., 1997. Functional rafts in cell membranes. *Nature* 387, 569-572.

Singer, S., 1972. A fluid lipid-globular protein mosaic model of membrane structure. In Membrane structure and its biological applications, *Annals of the New York Academy of Sciences,* New York, Green, D. (Ed.) 1972.

Singer, S., Nicolson, G., 1972. The fluid mosaic model of the structure of cell Membranes. *Science* 175, 720-731.

Swadlow, H., Waxman, S., 1976. Variations in conduction velocity and excitability following single and multiple impulses of visual callosal axons in the rabbit. *Exp. Neurol.* 53, 128-150.

Tanford, C., 1989. *Ben Franklin Stilled the Waves: An Informal History of Pouring Oil on Water with Reflections of the Ups and Downs of Scientific Life in General.* Duke University Press, Durham.

Tauc, L., Hughes G., 1963. Modes of initiation and propagation of spikes in the axon of molluscan central neurons. *J. Gen. Physiol.* 46, 533-549.

Tiwari-Woodruff, S., Schulteis, C., Mock, A., Papazian, D., 1997. Electrostatic interactions between transmembrane segments mediate folding of Shaker K^+channel subunits. *Biophys. J.* 72, 1489-1500.

Tomishige, M., Sako, Y., Kusumi, A., 1998. Regulation mechanism of the lateral diffusion of Band 3 in erythrocyte membranes by the membrane skeleton. *J. Cell Biol.* 142, 989-1000.

Tomishige, M., Kusumi, A., 1999. Compartmentalization of the erythrocyte membrane by the membrane skeleton: intercompartmental hop diffusion of Band 3. *Mol. Biol. Cell* 10, 2475-2479.

Torshin, I., and Harrison, R., 2001. Charge centers and formation of the protein folding core. *Proteins: Structure, Function, and Genetics* 43, 353-364.

Tsui-Pierchala, B., Encinas, M., Milbrandt, J., Johnson, E., 2002. Lipid rafts in neuronal signaling and function. *Trends in Neurosciences* 25, 412-417.

Vale, R., Banker, G., Hall, Z., 1992. The neuronal cytoskeleton. In: Hall, Z. (Ed.), *An Introduction to Molecular Neurobiology*. Sinauer Associates, Sunderland.

van Essen, J., 1973. The contribution of membrane hyperpolarization to adaptation and conduction block in sensory neurones of the leech. *J. Physiol.* 230, 509-534.

Vassilev, P., Dronzin, R., Vassileva, M., Georgiev, G., 1982. Parallel arrays of microtubules formed in electric and magnetic fields. *Biosci. Rep.* 2, 1025-1029.

Vassilev, P., Dronzin, R., Valevski, G., Kanazairska, M., 1983. In vitro polymerization of tubulin modified by application of low-intensity electric and magnetic fields. *Stud. Biophys.* 94, 139-140.

Vater, W., Stracke, R., Böhm, K., Speicher, C., Weber, P., Unger, E., 1998. *Behavior of individual microtubules and microtubule bundles in electric fields*. Presented paper at 6th Foresight Conference on Molecular Nanotechnology.

Vinogradova, O., 2001. Hippocampus as comparator: role of the two input and two output systems of the hippocampus in selection and registration of information. *Hippocampus* 11, 578-598.

Wallace, R., 1995. Microscopic computation in human brain evolution. *Behav. Sci*. 40, 133-158.

Wallace, R., 1996a. Microcomputational evolution of the neural membrane. *Nanobiology* 4, 25-37.

Wallace, R., 1996b. Quantum computation in the neural membrane: Implications for the evolution of consciousness. In: Hameroff, S., Kaszniak, A., Scott, A. (Eds.), *Toward a Science of Consciousness: The First Tucson Discussions and Debates.* The MIT Press, Cambridge.

Wallace, R., 1999. Computational aspects of neural membrane biophysics. In: Leszczynski, J. (Ed.), *Computational Molecular Biology*. Elsevier, New York.

Wallace, R., Price, H., 1999. Neuromolecular computing: a new approach to human brain evolution. *Biol. Cyber*. 81, 189-197.

Wallace, R., 2004. Neural membrane field effects in a cytoskeleton corral: microdomain regulation of impulse propagation. *Int. J. Quantum Chem*. 100, 1038-1046.

Wallace, R., 2007. Neural membrane microdomains as computational systems: toward molecular modeling in the study of neural disease. *BioSystems* 87, 20-30.

Wallis, C., Wenegieme, E., Babitch, J., 1992. Characterization of calcium binding to brain spectrin. *J. Biol. Chem*. 5, 4333-4337.

Wallis, C., Babitch, J., Wenegieme, E., 1993. Divalent cation binding to erythrocyte spectrin. *Biochemistry* 32, 5045-5050.

Wang, T-Y., Leventis, R., Silvius, J., 2000. Fluorescence-based evaluation of the partitioning of lipids and lipidated peptides into liquid-ordered lipid microdomains: A model for molecular partitioning into “lipid rafts”. *Biophys. J.* 79, 919-933.

Wang, T-Y., Leventis, R., Silvius, J., 2001. Partitioning of lapidated peptide sequences into liquid-ordered domains in model and biological membranes. *Biochemistry* 40, 13031-13040.

Whatley, V., Harris, R., 1996. The cytoskeleton and neurotransmitter receptors. *Int. Rev. Neurobiol.* 39, 113-143.

Williamson, I., Alvis, S., East, J., Lee, A.,, 2002. Interactions of phospholipids with the potassium channel KcsA. *Biophys. J.* 83(4), 2026-2038.

Winckler B., Forscher P., Mellman I., 1999. A diffusion barrier maintains distribution of membrane proteins in polarized neurons. *Nature* 397, 698-701.

Zajchowski, L., Robbins, S., 2002. Compartmentalized signaling in membrane Microdomains. *Eur. J. Biochem.* 269, 737-752.

Zuckermann, M., 1993. Self-sustained potential oscillations and the main phase transition of lipid bilayers. *Biophys. J.* 64, 1369-1370.

Index

A

B

C

D

K

L

M

N

O

P

T

U

V

W

X

Y